UNITED STATES
USJFCOM
JOINT FORCES COMMAND

Joint Operating Environment

Trends & Challenges for the Future Joint Force Through 2030

December 2007

The Joint Operating Environment document is intended to inform joint concept development and experimentation throughout the Department of Defense with a perspective on future trends, potential shocks, military implications, as well as the implications of these issues for the future joint force commander. This document is speculative in nature and does not necessarily represent a USJFCOM or Department of Defense position on any issue. Rather, it is intended to serve as an intellectual "starting line" for discussions about the structure of the future security environment. Inquiries about the JOE should be directed to USJFCOM Public Affairs, 1562 Mitscher Ave, Suite 200, Norfolk, VA 23551-2488. (757) 836-6555

Distribution Statement A: Approved For Public Release

Table of Contents

Chapter 1: Introduction

Joint and service concept development and experimentation applies innovation and change to the ideas and capabilities that animate U.S. military forces to ensure that they are adapted to the world in which they operate. The process of innovation and change relies on an ability to foresee and anticipate the key factors that shape the world and the challenges which are intrinsic to these factors. A forward looking, anticipatory view of change allows the United States to properly shape its forces so that they are prepared to deal with emerging challenges to our nation's security interests. Because the future is also shaped by shocks and other events that are by definition difficult or impossible to foresee, a culture of innovation and change encourages us to remain agile and able to quickly adapt in the face of surprise.

To encourage adaptation, the U.S. national defense and security community must at times take a step back and look beyond the imperatives of current operations. The Joint Operating Environment (JOE) document is an effort to take this longer view to better understand change in the international system. This view focuses on the period some 8-25 years in the future. As a first step, the U.S. defense and security community must forge a broad understanding of the key components of the international system that are relevant to the application of military power. Second, we must understand how these components evolve over time. Perhaps most important, we must understand the likely implications to military operations of these trends and shocks for the joint force.

The logic of trends and shocks will allow us to examine a number of models of potential future operating environments by combining different trends together to form plausible alternative futures. These futures may at times seem strange or unlikely – but an observer from 1980 provided with a description of the operating environment of 2007 would seem no less strange. The JOE document will articulate a set of military implications that reasonably flow from these wider trends in the international environment. Together, our discussion of trends, shocks, challenges, and military implications will be our "foothold in the future" for the wider joint and service experimentation community to consider when exploring new concepts, technologies, organizing principles, other methods to work in this future.

1.1 A Users Guide to the Future

The Joint Operating Environment document provides a framework for the study and articulation of a range of alternative future operating environments. The JOE presents future joint operating environments that have been developed after a wide-ranging examination of global, environmental, sociological, technological, and military dynamics that will influence the course of future conflict. The JOE document is intended to provide a research-based grounding for further discussions about the implications of potential future operational environmental trends for the joint training, experimentation, doctrinal development, and operational communities. These alternative futures can then be used to support the development of joint and service concepts, scenarios, experiments, exercises, and long term operational plans. By examining a number of critical trends influencing potential future operational environments and associated threats, this paper will serve as a common frame of reference and guide for civilian and military leaders responsible for the capabilities-based joint transformation process.

To support experimentation of future capabilities, the future must be viewed in a manner which accommodates a range of potential operating environments borne of these alternative futures, so the JOE will avoid presenting one particular vision of the future but will provide the reader the range of possible future trends and their potential implications for our military forces. Concept developers and experimenters should consider each of the trends described in this study, varying and adjusting the scale of trends in order to derive a specific military challenge or notional scenario.

1.2 Structure of the JOE Document

The remainder of the introductory section of the JOE document describes our view of the key elements that affect the international environment, the method by which the document is updated, and the JFCOM partners that support and contribute to this effort. The second chapter of the JOE document begins with a discussion of the key trends that will cause change in the future operating environment, including the scope, speed, and direction of changes to the joint operating environment over time. Chapter three presents a set of potential military challenges that are derived from the trends described in chapter two. Chapter four of the JOE document illustrates a set of military implications resulting from combinations of trends (and possible shocks) in the international environment that directly affect the application of military power. Thus, our discussion of trends, shocks, challenges, and military implications within the JOE document is meant to form the basis for debate and discussion for those responsible for exploring and shaping the future joint force.

Operating Environment, Trends, and Shocks

An international environment is composed of a large number of trends – some of which are relevant to military operations and some which are not. The JOE describes a set of trends that are relevant to national security and have the potential to cause conflict and war. This set of trends has been developed to define potential future operating environments and are enduring in nature – they have caused wars in the past and the present, and will continue to influence the course of human events in the future. Trends may be subject to shocks that accelerate or wholly change the direction of a trend. As trends and shocks accumulate in the operating environment over time they result in a new, future operating environment. This new operating environment has a set of military challenges and associated implications for the structures, capabilities, and functions of military forces. A more precise description of these important JOE terms follows:

- *The operating environment* is the combination of components, elements, factors, and other building blocks that describe the key features of the world in which past, current, or future joint forces will function.
- A *Variable* is simply something that varies from one state to another. Current and future operating environments are characterized by a set of variables. In this document variables allow us to characterize the differences between today's operating environment and tomorrow's, as well as differences among potential future operating environments. An example of a variable is the U.S. share of the world economy.
- *Trends* are the direction and speed of change in some important feature of the international environment. Trends document ongoing changes to these features or how they are accelerating or decelerating. Trends allow us to imagine possible

characteristics of a future operation environment. An example of a trend is that the U.S. share of the world economy is decreasing. Trends affecting the future operating environment are described in chapter two of this document.

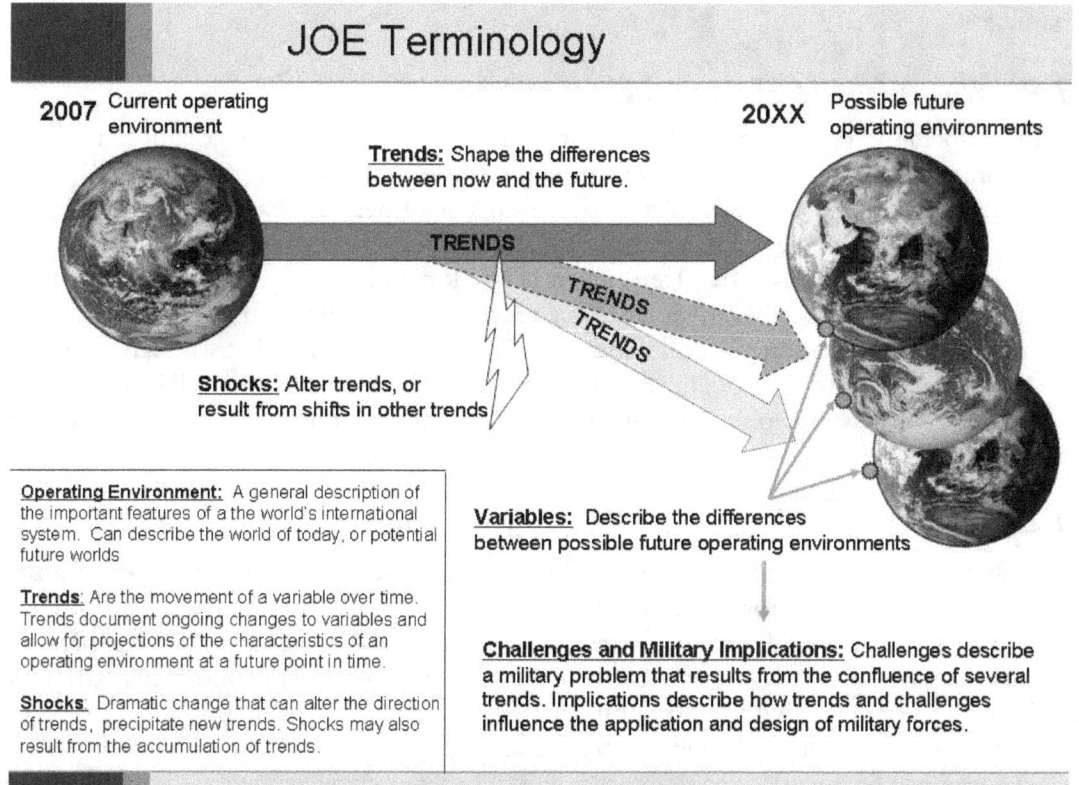

Figure 1.1. JOE Terminology

- **Shocks** are events that accelerate or decelerate a trend, reverse the direction of a trend, or even precipitate a new trend. Simply put, shocks alter the course of trends. Shocks can be sudden and violent, and are often unanticipated. They can also occur when a system passes a critical point and undergoes a "phase change." This type of shock results from the gradual accumulation of change in a number of variables (e.g. increased violence and frequency of hurricanes as a result of rising ocean temperatures).
- **Challenges** describe a specific military problem that is a result of one or more trends. The future joint force can expect to face a number of challenges based on the trends described in the JOE and are listed in Chapter 3 of this document.
- **Military Implications** relate trends to military capabilities. Implications describe in military terms why trends are important for concept developers and experimenters to consider in exploring and developing future joint force capabilities, and how trends might influence the conduct of military operations in the future.

A simple description of our consideration of the future might be written in this way: The current operating environment is composed of a set of variables which change over time. These trends can be further altered by events known as shocks. As trends and shocks accumulate over time a new operating environment – one that differs from the current environment – will emerge. Different trend projections and possible shocks may result in a range of alternative futures. The future operating environment described by these alternate futures has distinct military challenges and implications for the joint force commander that must be considered when developing new concepts, experiments, or strategies for future U.S. military forces.

1.3 Partners in the Futures Enterprise

The Joint Operating Environment document is the result of an ongoing collaborative research and exploration effort coordinated by the Deep Futures Division of USJFCOM's Joint Futures Lab. The observations recorded in this document are based on a cooperative effort between USJFCOM, and a variety of military, government, industry, and non-governmental organizations. The Deep Futures Division will work to stimulate an enduring dialogue that will foster further investigation and refinement. Each year, USJFCOM and its partners in the futures enterprise will conduct a series of workshops that will explore one or more of the trends within the document. The results of these workshops will be incorporated into new versions of the JOE which will then become the starting point for the next study effort.

1.4 A Look Ahead – Key JOE Observations

"The Future" is a very large and difficult to summarize topic. However, several key features relevant to the future joint force can be isolated and described in summary. The future trends portion of the JOE describes a world in which rich and prosperous states represent a smaller and smaller portion of humanity, while the poorest and least economically dynamic societies on earth grapple with rapid population growth and growing mega-cities, and cultural and environmental change that stresses already-fragile social and political structures. Globalization will lift millions from abject poverty, but its uneven impact will produce social dislocation, and because of raised expectations may produce dissonance and disorder if societies cannot translate gains in global trade into local prosperity. As more people around the world have access to markets, trade and travel, these flows become more vulnerable to disruption. Finally, greater complexity in the operating environment and rapid rates of technological change and surprise are changing security paradigms, placing greater emphasis on the prevention of conflict and blurring enemies, adversaries, competitors, and friends.

A number of joint force challenges will result as combinations of trends transform today's world into tomorrow's. These include enduring and emerging challenges, as well as national security shocks that may potentially change the international playing field in significant and possibly unpredictable ways. Enduring challenges for the future joint force will include familiar military activities, such as defending against attacks on U.S. territory, conflict with other powers, terrorist networks and criminal organizations. Enduring challenges also include dealing with the collapse of functioning states and the use of military forces to deter and prevent conflict around the world.

The joint force will also encounter a number of new and emerging challenges, the outlines of which are just becoming clear from our vantage in 2007. These include the development by states of anti-access strategies and capabilities, the potential emergence of new terrorist ideologies, and groups or states bent on the disruption of global trade and finance. The future joint force will likely be confronted by persistent cyber-conflict and the potential disruption of global information networks. The proliferation of weapons of mass destruction, failing nuclear and energy states or mega-cities will challenge the joint force to impose levels of order on highly disordered situations. A final emerging challenge is the potential growth and development of a global anti-American coalition of opportunistic states, transnational terrorist groups or supernational organizations.

Several national security shocks are identified that are clearly possible and should they come to pass, would have dramatic effects on U.S. national security and the wider global security environment. These include significant disruptions to energy security or conversely, the development of alternatives to oil. Other shocks include technological surprise, loss of access to the global commons or the emergence of man-made or natural pandemic that kills and sickens a significant portion of the world's population. Finally, nuclear attack on one or more of America's cities, or a global depression that disrupts the U.S. economy would overturn the international system and result in wide-ranging and dramatic changes to the U.S. security posture.

The JOE considers trends and military challenges and submits to the reader a family of military implications of the future operating environment for the joint force. These implications are arranged into five general categories -- terrain, base, knowledge, force application and command – that reflect key considerations of joint force commanders when planning for or conducting battles, strikes, and the overall campaign in support of national strategy. The joint force implications are not comprehensive, but are illustrative of the potential military capabilities that will be required in the future.

The JOE document concludes with a set of more specific military problem statements that are intended to stimulate discussion among the joint concept development, experimentation, planning, and operations communities to engage in a discussion of military problems inherent in the future operating environment and to encourage common exploration of potential solutions to them.

Chapter 2: Trends in the Joint Operating Environment

This section of the Joint Operating Environment Document explores a wide range of trends that have the greatest potential to shape or influence current and future military operations. An operating environment is an overarching term that encompasses a large number of more descriptive trends that influence the course and conduct of military operations. An understanding of the operating environment is central to our ability to engage in and win any conflict. This section of the JOE document provides a framework for considering trends in the future operating environment that will influence joint force operations with a focus on the environment that the joint force can expect to encounter some 8 to 25 years from today.

A number of trends occurring within this landscape characterize ongoing changes in the operating environment and assist in our ability to think about the future geopolitical landscape and how that landscape will influence military operations. While trends often change slowly over time they can dramatically accelerate or be altered by unanticipated shocks. As examples, demographics tend to change relatively slowly, while the rate of technological change can be quite rapid. As trends and shocks accumulate over time a new operating environment – one that differs from the current environment – will emerge. This new *future* operating environment has military implications that must be considered when developing new concepts, experiments, and strategies for future U.S. military forces. Trends in the joint operating environment are arranged into four general categories based on similarity in order to help the reader better comprehend them. These categories are:

- Human Geography.
- Governance and Legitimacy.
- Globalization of Economics and Resources.
- Science, Technology, and Engineering.

The trends described in this section of the JOE address the likely context of future military operations within an environment characterized by a global and interdependent world with massive human and technological change. The trends that follow will support our ability to derive strategic and operational implications for the future joint force and provide context regarding the character and form of future conflict and war.

2.1 Human Geography

This trend category describes the quantity, characteristics, and distribution of human populations. It includes factors such as where humans live, work, how they move, and how their characteristics change over time. Human geography also involves how groups of human beings relate to the natural environment and how their activities affect the ecology around them. Although some changes at the margins may occur – beachfront erosion, volcanic activity, population migration from drought, for example – it is how those changes affect groups of people that will be our subjects for consideration in this paper. The following trends in the geography of humanity will be a defining feature the future operational environment that the joint forces will be required to consider in its planning and operations.

Population Growth

The last decade has seen the more aggressive predictions of human population growth not coming true, as the overall trend is toward lower rates of growth and fewer births per mother, especially in Europe, Japan, Russia, and China. In 1989, the world was adding 87 million people per year; by 2002 the growth had dropped to 74 million additional people every year. Most of the decrease in the gross number of births has been traced to a general decline in the number of children born per woman, which has dropped from 3.05 births per mother in 1990 to 2.55 per mother in 2005.

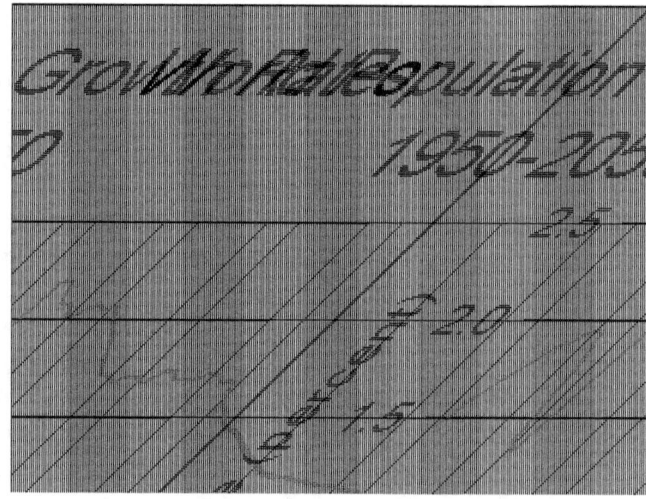

Figure 2.1. Growth Rate: World Population

Regardless, humanity continues to add people at a fairly staggering rate, and the resulting demographic trends have significant implications for the future security environment. The US Census Bureau population clock estimates that world population will be 6.9 billion people by 2010.[1] Most analysts agree that the world's population will continue to increase, possibly to 8.3 billion by 2030. The impact of changing demographic varies greatly across the globe. The primary implication will be massive human movements as stressed populations in poor nations seek a better life elsewhere. This trend will continue the massive increase in urban populations and create political and social strains within and among developed and developing nations. In groups that do not move, growing populations will increase the urgency of the competition for resources, notably food, water, and energy. Such shortages, combined with other tensions in these areas, could be a significant trigger for conflict.

More than 90 percent of this increase will occur in developing and poorer countries.[2] While the global population rate will increase by 3.7 percent from 2010 to 2030, the African population will grow by 7.7 percent, while the population of the very least developed countries (some in Africa, others in Southwest Asia) will increase by 8.2 percent. The countries most affected by future population growth trends will likely include India, Pakistan, Nigeria, China, Bangladesh, Ethiopia, and the Democratic Republic of Congo. Regionally, countries in North Africa and the Middle East are also predicted to experience rapid population increases. These countries will not be able to meet many of the basic needs of their expanding populations.

In developed areas such as Western Europe, nations will be challenged to maintain economic productivity, to provide the promised level of social support to their aging populations, and to provide enough manpower for their armed forces. Current projections show that by 2030, Europe's population will decrease by 4.7 million. The only possibility of reversing this population loss – immigration – has enormous impact for the political, social,

---

[1] United States Census Bureau, <u>World Population Clock Projection</u>.
[2] The United Nations, *World Urbanization Prospects: The 2005 Revision*.

and cultural makeup of Europe. The primary source of most immigrants is North Africa and the Middle East. There is already great tension between native born and immigrant Europeans.

Japan faces many of the same challenges but the current social contract in Japan prevents any significant immigration to alleviate the shortage of labor. Japan may be only the first to invest very heavily in robots in an effort to provide the basic physical labor needed by the society. Russia has seen a major withdrawal of it population from the Far East and Siberia. Chinese have migrated into the void and Russian nationalists have begun demanding action to prevent a Chinese takeover of Russia's Far East. China itself will face demographic pressures as it population ages. Due to its one child policy, China will be first country to grow old before getting rich because its birth rate remains below the replacement rate of 2.1 children per woman. China faces the 1-2-4 problem with four grandparents having two children and one grandchild – a demographic profile that makes intergenerational pension programs very difficult to support.

Age Distribution

It is noteworthy that the proportion of the population of working age (15 to 59) is expected to decrease between 2005 and 2050 in every major area except Africa. This reduction is part of the process whereby the beneficial ratio of workers to dependants starts decreasing in all major areas except Africa. In fact, the support ratio, calculated as the ratio of persons aged 15 to 64 over the sum of the number of children and of persons aged 65 or over, is expected to begin declining after 2010 in Europe, Northern America and Oceania, after 2015 in Asia and after 2025 in Latin America and the Caribbean. By 2050, Europe is expected to have the lowest support ratio, at 14 persons of working age for every 10 dependants. Other regions are expected to have support ratios ranging from 16 in Northern America to 19 in Africa. Although the oldest populations in the world are found in developed countries, 64 percent of the older persons alive in 2005 lived in developing countries and by 2050 nearly 80 percent of those aged 60 years or over are expected to live in developing countries.[3]

Increased life expectancy and falling fertility rates will contribute to a continuing shift toward an aging population in the most developed countries. By 2030 close to 1.4 billion people in the world will be over the age of 60.[4] Medical science and better public health programs are prolonging life and already creating an increasingly older population in Europe and the United States. The declining ratio of working people to retirees in "aging" developed countries will strain already taxed social services, pensions, and health care. China is also part of this trend. By 2030, China will have 348 million citizens over the age of 60, nearly as many people over age 60 as the projected total population of the United States.

Conversely, people in poor countries continue to have large families for many reasons. Prominent among them is little or no access to family planning services. In the past, many of these children would have died, but now medical science has saved them—creating a youth bulge of undereducated youth with few employment opportunities but with heightened expectations. Globalized information networks will tend to raise expectations in

---

[3] Mark Haas, "A Geriatric Peace? The Future of US Power in a World of Aging Populations," *International Security*, Summer 2007, pp. 112-147.

[4] United Nations Department of Economic and Social Affairs. *World Population Prospects: 2004 Revision.*

poorer societies, creating envy and mistrust of more prosperous states, while population age differences generate differing perspectives on both problems and solutions.

It will be very difficult for regional labor markets and economies to absorb this forthcoming "youth bulge."[5] (15 to 29-year-olds) that will occur in Sub-Saharan Africa, Latin America, and the Middle East. People in the 15-to-29 age group place significant demands on governments and society. Without education and opportunity, the higher the density of this age group (approaching 40-50 percent of a population) the greater the potential for instability of the state.[6] During the last two decades, 80% of all civil conflicts occurred in countries in which 60% or more of the population was under the age of thirty.

This youth bulge provides fertile ground for recruits in terrorist groups, criminal elements, and drug cartels. Historically, if this segment of the population is unable to find adequate education or employment and if expectations go unmet, then social chaos is inevitable and has the potential of turning into conflict. If this holds true in the future, the resulting disparity between aging developed countries and countries with young, undereducated, underemployed populations will exacerbate the frustration of the less fortunate, who will understand the benefits associated with globalization, but will not share in those benefits.

Sex Ratio

In many societies where traditional cultures have access to improved pre-natal medical techniques couples have preferred male children to females, which is reinforced by government policies limiting families to fewer (or even one) children. This sex selection tendency has led to a growing imbalance in the ratio of males to females as a result of sex-selective abortion and infanticide (notably in China and in India). Furthermore, large-scale immigration of male laborers unable or unwilling to bring their families with them (as in many Persian Gulf countries), will fuel social unrest in their native countries as the excess of young males will be unable to find spouses.

Climate Change

Earth's climate is the macro-component of the physical environment, and a critical component of the future operational environment. Significant changes will have effects across all of the variables. In the past, oscillations in temperature cycles have caused plunging agricultural yields, leading to famine, rebellion, and war, and over the longer term, dynastic collapses. While the climate has historically varied between hot and cold periods, recent projections focus on the potential effects of climate change due to "global warming." A recent report states that there is a more than 90 percent probability that significant increases in global average temperature over the last 50 years are related to human activity. The report further states that these increases are causing worldwide effects that have the potential to continue well into the future.[7]

---

[5] De Benitez, Sarah Thomas et al. (2003) *Youth Explosion in Developing World Cities: Approaches to reducing Poverty and Conflict in an Urban Age.* (Washington, D.C.: Woodrow Wilson International Center for Scholars). page 12.
[6] Deputy Chief of Staff for Intelligence, US Army Training and Doctrine Command. Sociology panel, Joint Operational Environment Seminar. Williamsburg, VA, June 2003.
[7] A Report of Working Group I of the Intergovernmental Panel on Climate Change: Summary for Policymakers. p. 10.

The predicted effects of climate change over the coming decades include extreme weather events, drought, flooding, sea level rise, retreating glaciers, habitat shifts, and the increased spread of life-threatening diseases. The World Health Organization estimates that climate change is already responsible for an estimated 150,000 deaths per year. These predicted effects are likely to vary regionally, within varying ranges of change and severity and speed of onset. The potential of these effects serve as a kind of "wild card" within the JOE as some effects may not be recognized until after the fact, and by 2025-2030, they may or may not be causing widespread recognized effects. Despite the uncertainty regarding effects, the potential for unpredictable rapid change exists.

Projected climate change will seriously exacerbate already marginal living standards in many Asian, African, and Middle Eastern nations, causing widespread political instability and the likelihood of failed states. Unlike most conventional security threats characterized by the activities of single entities acting in specific ways, climate change has the potential to result in multiple chronic conditions, occurring globally within the same time frame. Economic and environmental conditions in already fragile areas will further erode as food production declines, diseases increase, clean water becomes increasingly scarce and large populations move in search of resources. Weakened and failing governments, with an already thin margin for survival, foster the conditions for internal conflicts, extremism, and movement toward increased authoritarianism and radical ideologies.[8]

The U.S. may be drawn more frequently into these situations, either alone or with allies, to help provide stability before conditions worsen and are exploited by extremists. The U.S. may also be called upon to undertake stability and reconstruction efforts once a conflict has begun, to avert further disaster and reconstitute a stable environment. Effects may spread to the U.S. Homeland in the form of refugee flows, internal weather-related disasters, energy crises, and associated terrorist activities. Potential strategic implications may include the potential opening of new sea lanes and access to new resources as a result of the melting Arctic ice cap and tensions regarding availability or reallocation of energy resources. Climate change may also have impacts on areas of military capability ranging from trafficability, to potential inundation of military ports and other bases to sensor performance.

Crime

Any future operational environment will include the presence of criminal elements. International organized crime, motivated by greed and self-interest, may increase as potential security threats to the developed world with rising numbers of young people and governance and law enforcement mechanisms not able to evolve rapidly enough to suppress new forms of crime.[9] Along with a rise in the number and presence of criminal organizations, there will also be an increased blurring of criminal activities, civil conflict, and potential terrorist activities (see Figure below).[10] These elements will continue to blend with the population and may become ever more difficult to penetrate as criminal networks become more sophisticated and capable. Drug and human trafficking are expected to continue. Such organizations and activities will threaten national or regional stability, structure, and legitimate political authority. This, in turn, can affect U.S. interests. Criminal organizations

---

[8] National Security and the Threat of Climate Change, (The CNA Corporation, 2007).

[9] United Nations Office on Drugs and Crime. *Annual Report 2005.*

[10] Figure from International Crime Threat Assessment (December 2000).

and elements will take advantage of information and communication technologies and the proliferation of weapons to develop very sophisticated capabilities. The destructive social, economic, and political impact of crime will increase in both its severity and sophistication.

Transnational criminal activity, fueled by global connections to money and arms, will blur the lines between traditional military action and criminal activities. Criminal organizations will continue to form strategic alliances with states and non-state actors, including terrorists. Terrorists and criminals will also be active in such an environment, ready to exploit the situation for their respective gains. United States joint forces, combined with law enforcement and intelligence activities in a collaborative information

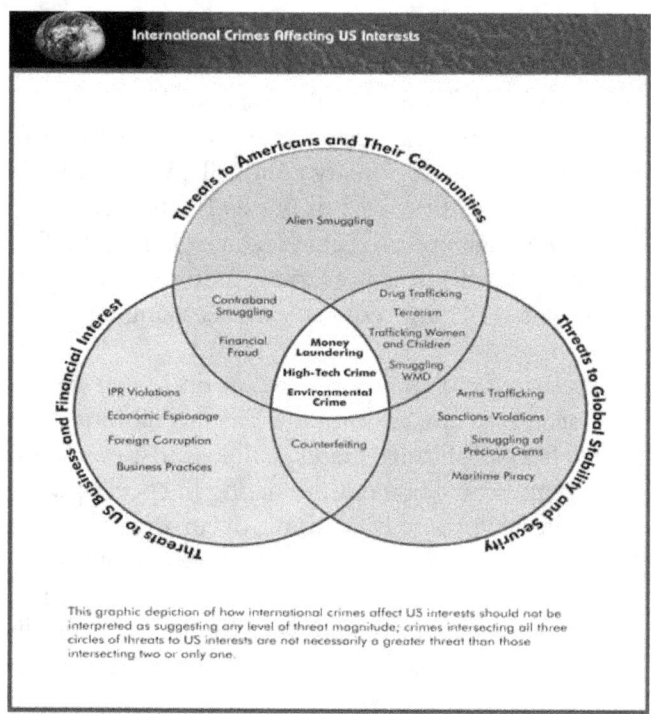

Figure 2.2. International Crime

environment, will have to deal both with enemy military forces and other non-traditional forces, such as criminal organizations, terrorists, or religious fanatics, who will seek to profit from instability.

Culture, Faith, and Ethnicity

Culture will remain a source of friction and potential conflict among societies. The future operational environment must accommodate a significant trend in the growing significance of culture, cultural differences and sub-cultures as a force for conflict. Fed by globalization, regionalization, and information age capabilities, new groups are discovering (and sometimes rediscovering) a shared culture. This trend complicates our ability to define, understand, and influence the future operational environment.

Religion will remain an important force in the joint operating environment. Religion is an aspect of culture that frequently has created friction and conflict. Where communism and fascism were once used to motivate oppressed, impoverished, or culturally adrift populations, peoples seeking national, regional, or even global goals of dominance will increasingly employ religion, particularly in an extreme and often violent form. Indeed, religion has already taken its place with neo-nationalism and racial solidarity as ideological pretexts for uniting peoples and justifying conflict. Religious radicalism and fundamentalism will become attractive to those who feel victimized or threatened by the cultural and economic impacts of globalization and increased social interconnectivity. The rise of radicalism, especially religious radicalism, will complicate any strategic action from all elements of power, altering the character and form of combat.[11] This notion changes the

---

[11] Barnett, Thomas P.M. "The Pentagon's New Map," *Esquire Magazine.* March 2003.

calculus of conflict in which the stakes become higher, supreme sacrifice is more prevalent, and the perspectives on the constituent elements of will are different. Moreover, religious radicalism may be transnational and greatly empowered by global information systems that allow participation, recruitment, planning, collaboration, and resourcing, regardless of borders or states.

The access to and awareness of other cultures, either through direct contact with individuals or through technologies that spread the popular culture of other groups will increase dramatically by 2020-2030. Western cultural influences, enabled by information technologies, have and will continue to have worldwide impact. How a country or a people, respond to the influence of and encounters with other cultures will be an important element in future conflict, and may replace other more traditional internal drivers as a key source of tension.[12] In general, globalization is viewed positively, especially in poorer countries as opposed to richer, developed countries.[13] Globalization will continue to increase the intensity and breadth of outside influence on all cultures.[14]

Issues of ethnic cohesiveness and ethnic tension will be important to military forces operating in any future operating environment. Similar to cultural conflicts, ethnic conflicts tend to rise when identities are challenged by the kinds of major social changes that accompany modernization, globalization and migration. While current ethnic fault lines tend to be geographically centered, in the next twenty years globalization will likely unite ethnic Diasporas around the world, including within the United States itself. In Indonesia, for example, there are seven million ethnic Chinese who are able to influence the economy back home in China through remittances; that is, having amassed great wealth, these migrants regularly send money home to their families in China. In the future, therefore, understanding and recognizing the ethnic makeup of a given environment and its ties to a global community will be even more critical.

Intercultural encounters are not always positive. In very underdeveloped countries with a large, unemployed youth bulge, Western cultural influence results in disaffection and resentment—both of which fuel crime, terrorism, and drug usage. External cultural infusion leads to a weakening of cultural cohesiveness, producing a backlash of negative attitudes and actions. These attitudes contribute to the development and spread of intense anti-Western sentiment that presents major political challenges or increased terrorist acts against U.S. interests and personnel.[15] The ongoing fundamentalist reaction to the perceived cultural domination by the U.S. inevitably led to anti-U.S. sentiments and acts tailored to cripple the strategic power and image of the U.S. The U.S. political and military response to these acts provides fodder to the fundamentalist mindset, ensuring that these dynamics will continue into the foreseeable future.

---

[12] TRADOC DCSINT. Findings of Winter JOE Conference on Culture, (25-27 January 2005). Published 22 June 2005.

[13] David Dollar, "The Poor Like Globalization," within *YaleGlobal, June 23, 2003*.

[14] Joint Doctrine and Concepts Centre (JDCC). *Strategic Trends: The Social Dimension.* March 2003.

[15] Lonnie Henley, *Factors Sustaining Terrorism and Extremist Violence*, Defense Analysis Report, (Washington, D.C.: Defense Intelligence Agency, 17 February 2004).

Education

Trends in training and education are important indicators of the stability, productivity, and strength of a society. Education raises the potential for economic prosperity and political activity. Education is "the foundation for development and a future place in the global economy."[16] Education will become easier as it moves online, allowing far greater access to knowledge than ever before.

Education, however, is a function of the ruling body and its inherent bias. There is a global appreciation for the strength and quality of the U.S. university-level system and advanced studies programs. More than 30 percent of all science and engineering doctoral degrees awarded in the U.S. during academic year 2002-2003 went to nonresident aliens, with the majority of degrees in the areas of science and technology.[17] Over time this will tend to erode U.S. technical and scientific leadership. This erosion can already be seen in the case of India, which is now one of the world leaders in computer software development. Another trend is outsourcing of U.S. research and development and back-office and other information technology (IT) support; whether this outsourcing is positive or negative remains to be seen.

Further, education will raise awareness in the developing and underdeveloped world that one's own society's standard of living is far below that in the developed nations. The resulting gap between the haves and have-nots will create tension, especially if people can be educated, but there are no jobs. A corollary effect will be the loss of many technically trained professionals to more developed countries, with resulting impacts in many areas, including health care.

Despite the general growth in educational infrastructure worldwide, in many countries access to basic education continues to be denied to certain segments of society. The issue of fundamental education remains a concern in many parts of the world, and the education gap is widening further. As of 2000, there were at least 880 million illiterate adults globally; 250 million children worked, and more than 110 million school-age children did not attend school.[18] These figures represent more than 1 billion people inadequately educated to participate in or benefit from the growth of the global economy. Providing adequate training and education to compete successfully in a highly technical global environment is an obligation of society. If a segment of a population feels marginalized; unable to compete for jobs; educated but without outlets for such education; or denied basic education, it is likely that civil strife and violence, or even revolution may occur.

Health

Although advances in health services have been made worldwide during the last decade, many people still lack access to basic medical care and treatment. It is expected that disparities between health care in the developed and developing world will widen. Biotechnology is likely to further increase the gap between haves and have-nots by extending the life-span and quality of life of the rich. Chronic and infectious diseases will continue to

---

[16] Kofi Annan, Speech to World Education Forum, Dakar, Senegal, (April 2000). http://srch1.un.org.

[17] National Science Foundation, http://www.nsf.gov/statistics/nsf05300/tables/tab3.xls January 2005.

[18] UNESCO *Education for All Global Monitoring Report 2005*.

have a dramatic economic and social impact in Africa and parts of Asia and South America, causing more resources to be dedicated to fighting these diseases, and leaving less money for other basic needs. Infectious diseases including HIV/AIDS, malaria, hepatitis, and tuberculosis will be present in most future operational environments. Waterborne diseases account for about 90 percent of infections in developing countries where 95 percent of urban sewage is dumped untreated in rivers and lakes. Water pollution is estimated to cause fourteen thousand deaths each day worldwide. Increased urbanization will increase health problems caused by pollution and lack of sanitation. Increasingly, upstream pollution will be a source of conflict with downstream users.

Environmental change can change the established relationship between people and pathogens, facilitating new disease outbreaks. Climate change, for instance is expected to alter temperature and rainfall patterns, thus permitting tropical diseases to thrive in previously cooler areas where they could not survive before.

Urbanization

The twentieth century witnessed the rapid urbanization of the world's population. The global proportion of urban population increased from a mere 13 per cent in 1900 to 29 per cent in 1950 and, according to the 2005 revision of the U.N.'s *World Urbanization Prospects*, reached 49 per cent in 2005. Since the world is projected to continue to urbanize, 60 per cent of the global population is expected to live in cities by 2030. The rising numbers of urban dwellers give the best indication of the scale of these unprecedented trends: the urban population increased from 220 million in 1900 to 732 million in 1950, and is estimated to have reached 3.2 billion in 2005, thus more than quadrupling since 1950. According to the latest United Nations population projections, 4.9 billion people are expected to be urban dwellers in 2030.[19] Tokyo will likely remain the world's largest city, with Mumbai and Delhi following closely behind.[20]

Along with the greater economic opportunity cities provide, uncontrolled urban growth also contributes greatly to increased crime and instability. As people move to the cities, they tend to lose the traditional influences and patterns characteristic of small towns and villages. They are often forced to move to poor areas that are likely to provide minimal services and opportunity and be controlled by criminal gangs. Sprawling slums are growing in the megacities of Lagos, Karachi, Mumbai, Shanghai, Sao Paulo and others. These slums breed disease, poverty, criminality, and political discontent as demonstrated recently by powerful criminal organizations in Rio de Janeiro challenging government control and jurisdiction in the very heart of the country.

The last twenty years have seen a massive move of populations to urban areas worldwide. By 2015 there will be 22 mega cities (up from 20 today) some 17 of which are in the less developed world. The rate of growth in these massive urban areas will slow. However, the number and size of cities of between one and 10 million will continue to expand. Much of the expansion of large and mega-cities will take place in littoral, coastal regions within 100 kilometers of the coast. Some 60 percent of the world's population currently lives within

---

[19] The United Nations, Department of Economic and Social Affairs. *World Urbanization Prospects: The 2005 Edition*. Working Paper No. ESA/P/WP/200.
[20] *Ibid.*

100 kilometers of the ocean. Some 70 percent lives within 320 kilometers, and most cities with populations of more than one million are located in the littorals.[21] Furthermore, littoral regions include straits and other strategic chokepoints which are often surrounded by highly populated cities (such as the Strait of Malacca) from which the world's sea lanes of communications can be controlled. [22]

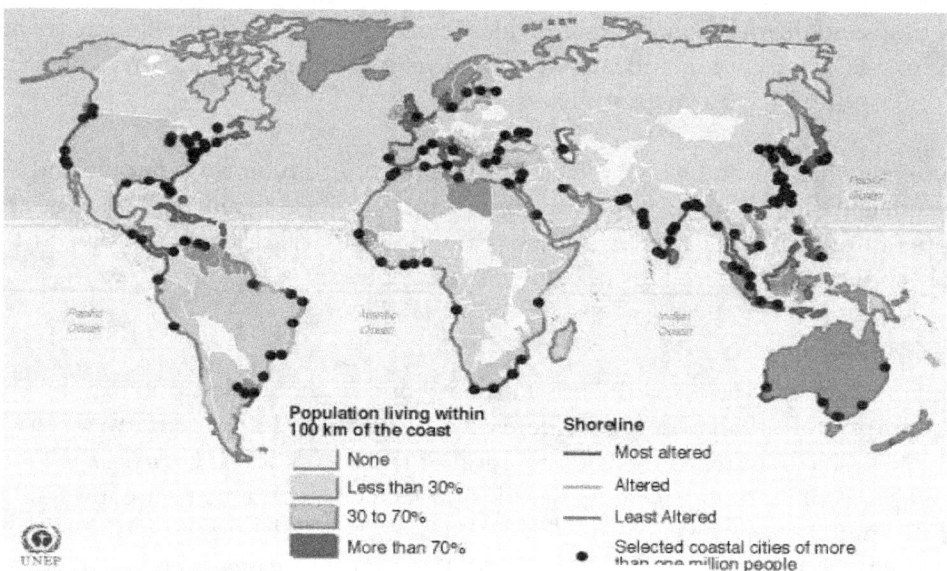

Figure 2.3. Urbanization of the Coast

Providing for these growing urban populations will challenge local governments and public service infrastructures. Many of the world's largest cities are in the developing countries. If their governments are unable to provide the basic public services, the potential for chaos and civil unrest will be heightened. On the other hand, if governments marshal resources to deal with urban issues, they risk leaving rural areas under- or ungoverned, with all the potential problems that could then develop. The critical infrastructure in such areas circa 2030 most likely will be austere—water and sewer services in disrepair; limited or compromised electrical service; inadequate medical care—directly affecting the American and coalition means to respond with military forces or humanitarian aid.

Migration

Urbanization is one form of migration and will absorb the vast majority of the economic migrants around the world. The other major movement of human populations will be the migration of people from poor countries and regions to wealthy ones. This movement will challenge the legal processes developed countries have placed on immigration. The US, Europe, and the Russian Far East have and will continue to face very large scale illegal migration. Each will have to develop security strategies that deal with this massive movement of humanity.

---

[21] *Marine Corps Midrange Threat Estimate—1997–2007: Finding Order in Chaos* (Quantico, VA: Marine Corps Intelligence Activity, August 1996) p. 1.

[22] Littoral Urbanization graphic from Burke, *et. al.* World Resources Institute, Washington D.C. *AASS Atlas of Population and Environment 2001*. (American Association for the Advancement of Science, University of California Press: Berkeley).

People generally migrate to those states and regions that can provide a better quality of life. Migration can have positive effects as developed nations receive needed labor and developing nations lose populations they cannot support. However, an inability to integrate these migrants into the economy may lead to severe economic isolation and alienation that can result in civil unrest. Furthermore, developed states will take the best and brightest of the developed world, encouraging a "brain drain" of skilled persons needed for the economic health of the less-developed world.

The effects of climate change, including droughts, floods and, potentially, rising sea levels could also contribute to increased migration. Large scale migration will be one of the major security issues related to climate change, primarily due to changes in food and water availability or proximate physical changes to their former locale, such as sea-level rise, desertification and fires, or forced relocation by security forces. Climate change may force migrations of workers due to economic conditions, and the movement of asylum seekers and refugees. Migrations in themselves do not necessarily have negative effects, although taken in the context of global climate change a net benefit is highly unlikely. Migration patterns may occur within countries, across borders, and across entire regions, and each type of migration brings different stresses relating to additional competition for diminishing available resources, increased demands on systems, infrastructure, racial and religious tensions and increased cultural, political and economic stress. [23]

The effects of labor migration, issues of migrant integration, and irregular migration flows can be further complicated by illegal activities such as human trafficking and the smuggling of migrants. All these things will contribute to potential flash points of any future operational environment. For example, developing nations will struggle to retain skilled and professional work forces while allowing unskilled and burdensome elements to migrate. Unfortunately, many immigrants will lack the skills necessary to compete in the middle-class work setting of the developed nations to which they migrate. Thus, many unskilled immigrants will become disenfranchised minorities with little voice in their new countries, competing with native populations for a diminishing number of unskilled labor positions.

Advances in communications and transportation will impact future migration, creating "virtual" borders and causing migration to be less influenced by physical proximity. Instant communications—visions of a better life, carried by radio, television, and the Internet —will be very influential. The cultural, economic, and historical impact of a multi-ethnic labor force could affect foreign policy or military operations. For example, with its large and growing Hispanic minority, the U.S. might either hesitate or be pressured into conducting major military operations in Latin or South America. Strategic alliances, partnerships, and coalitions may also be affected by the changing dynamics of member populations. Such factors will impact most strategic decisions, be they political, informational, military, or economic.

2.2 Governance and Legitimacy

The governance and legitimacy category encompasses a number of trends that relate to the nature of actors in the international environment, including who owns or can make use of

---

[23] National Security and the Threat of Climate Change, (CNA Corporation, 2007).

natural and human resources and who can command loyalty and claim legitimacy to act throughout the world. Today, actors in the international environment may consist of civilizations or religions that consist of billions of individuals or may be small groups or even a single person that can affect the course of history. Today, there is a wide and diverse array of actors capable of exercising power in some way on the international stage, and the traditional notion of the sovereignty and primacy of the nation-state is under challenge. Some states are delegating parts of their sovereignty upwards to international and supranational organizations, while aspects of sovereignty are draining downward to subnational identity groups.

The Powerful

The state will be redefined and perhaps weakened in the next 20 years; although it will remain a key element of international and regional relations. In affluent and developed countries consolidation of some traditional governance functions under regional economic and political supranational bodies, like the European Union (EU), will continue, but the member nation-states will retain significant powers. For example in Europe, states such as Belarus and Russia may join together to form a larger state or union. Regional supranational organizations represent a pooling of sovereignty to achieve greater collective power. As members, nations have the potential to become increasingly powerful, capable of concerted diplomatic, information, military, and economic actions. These organizations can constrain or facilitate America's ability to act or react.

Further redefinition of the state will be evidenced as political identities blur and some power shifts to nontraditional actors. For example, some international or private business organizations will proliferate and assume some of the powers now held by local and national governments. Others may become obsolete and dissolve, combine into different entities, or fade into obscurity because they are no longer relevant. States will find it increasingly difficult to act unilaterally and will have to be more adept at forming temporary alliances and multilateral arrangements. Otherwise, they will find themselves increasingly isolated, especially when entities in economic zones such as Europe or the Far East emerge with power equal to or greater than that of states. States experiencing significant economic growth will tend to develop pluralistic and liberal societies, as development and some form of democracy are usually interdependent. (It remains to be seen whether China will follow this tendency.)

Within economically successful states, identities, functions and allegiances of individuals, corporations, governments, and NGOs will dynamically change, blend, and disband as the information revolution, globalization, and international travel evolve. National identities and the identification with states may have less importance as dual citizenship becomes more common and as supranational organizations mature and multinational corporations spread. Allegiances will be tied more closely to cultural, religious, or ideological proclivities as stateless nations attempt to acquire one of their own. The Internet will enable interest groups to come together, morph, and disband with unprecedented speed. Cultures will merge, the English language will dominate, and brand-name products will become increasingly universal. International organizations and special-interest groups, and far-flung Diasporas will make competing claims on the loyalty and allegiance of their members because the Internet and global communications will strengthen the ties and power of scattered peoples. People will increasingly make decisions from data that generally leads to an inclination to make quick judgments and intellectual snapshots with no hint of knowledge, understanding, or complexity. The power of the media will be enormous.

The Weak

Many trends, particularly economic trends, that are driving developed nations do not influence less developed states to the same degree. Economic trends tend to weaken traditional religious, ethnic and cultural ties. Social and class mobility provide increased opportunity and political/cultural interaction. Without increased social and political advancement, groups and nations will be subject to arrested development and exploitation by non-state actors. In these areas, tribal and religious identification may supplant the formal governing structure and become the de facto government at the local or regional level. Potentially, less developed countries may become dictatorships, countries with warring internal entities or become ethno-religious states headed by clerics and ruled by religious doctrine or law.

Legitimacy

Regimes generally require the assent of a large proportion of the population in order to retain power; however, they may survive high degrees of unpopularity if they are supported by a small but influential elite group. The globalization of information and trade will complicate the perception of legitimacy as populations become increasingly aware of their status relative to other similar countries. Some governments will lose legitimacy and fail based on loss of public confidence and/or dissatisfaction with status as "have-nots."

Safety and security will be increasingly important, and individuals, groups, corporations, and governments will pay a premium to have it – increasing the number of "legitimate" security providers throughout the world. Security will be an expanding business, as corporations and governments operate in high-risk areas with a concomitant expansion of security risks. Safety concerns will override some human rights, liberties, and privacy. Industrial espionage and the advent of legitimate business intelligence will increase the requirement for personnel, physical, and electronic security. Military forces will have to expend additional resources on securing and protecting computer network operations (government and corporate) and force protection, and train as partners or members of a coalition with a wider variety of military, paramilitary, police, and government and private security forces. How citizens of various regions of the world accept or reject non-state actors using force will be an important consideration for future joint force planners and operators.

Failed or Failing States

Failed or failing states will arise as a result of economic collapse, resource competition, ideologically centered mismanagement, and failed social infrastructure. (See Figure 2.4 below) Some states or regions (for example, North Korea and Central Africa) will depend on foreign aid and handouts for survival. As a result, aid-dispensing international organizations or multinational corporations (MNCs) may provide de facto governance. Some areas that are currently ungoverned or lack effective government control (Northwest Frontier Province in Pakistan, Somalia, large areas of tropical Africa and South America) are at risk of remaining ungoverned and such areas may be an increasing feature of the international security environment. In these areas, local warlords, criminal bosses, tribal leaders, and religious authorities will rule. These areas will have increasing importance to desperate or disenfranchised citizens, while providing sanctuary for terrorists, criminals, and revolutionaries. Terrorists, drug dealers, and criminal elements will thrive in these sanctuaries and will use them as the base to spread their influence. They will migrate and return when necessary.

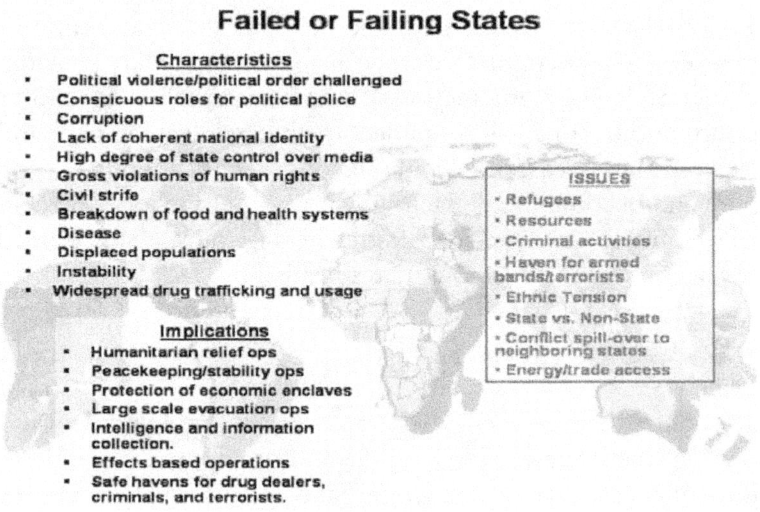

Failed or Failing States

Characteristics
- Political violence/political order challenged
- Conspicuous roles for political police
- Corruption
- Lack of coherent national identity
- High degree of state control over media
- Gross violations of human rights
- Civil strife
- Breakdown of food and health systems
- Disease
- Displaced populations
- Instability
- Widespread drug trafficking and usage

ISSUES
- Refugees
- Resources
- Criminal activities
- Haven for armed bands/terrorists
- Ethnic Tension
- State vs. Non-State
- Conflict spill-over to neighboring states
- Energy/trade access

Implications
- Humanitarian relief ops
- Peacekeeping/stability ops
- Protection of economic enclaves
- Large scale evacuation ops
- Intelligence and information collection.
- Effects based operations
- Safe havens for drug dealers, criminals, and terrorists.

Figure 2.4. Failed or Failing States

Transnationalism

Transnationalism is a trend that features increased physical and cultural connectivity among different peoples; is marked by large flows of people, ideas, and goods among regions and results in the marked decrease in personal attachment of individuals to the state. Transnationalism will have a profound effect on governance for both the powerful and the weak. It will significantly impact, either positively or negatively, how existing governments and future governments evolve. The impact may be direct (through political actions of dominant supernational organizations) or indirect (through the influence of global culture or global economy). Governments and regional alliances may come to view transnationalism as a challenge to their rights and legitimacy and will actively combat or subvert it. Transnationalism is eroding the ability of states to dominate particular issues as it increases the importance of transnational organizations to act. Thus the traditional application of the instruments of national power, such as diplomatic relations between states may also have to include negotiation and relations with transnational organizations.

The information revolution will effect change in governance worldwide. It can serve governance by empowering it through improved communication and education, likely resulting in positive effects overall. Proliferation of information will cause instability within those governments that attempt to isolate their populaces' awareness of their status as "have nots" relative to many other nation-states. The information revolution can also serve political revolution because it facilitates the identification and organization of like-minded persons around the world. Information technology allows them to form special interest groups, unite global or regional diasporas and share vast amounts of information quickly. Those able to afford the relatively low cost of information management technologies will have spectacular advantages over those who do not. For example, by exploiting such capability, friend, foe, and neutral will attempt to create political and economic influences that, taken as a whole, can cascade across national and organizational boundaries with immense effect.

Information technology will continue to spread world opinion that can influence and limit the power of government. It raised international attention and pressure against *apartheid* politics in South Africa and fueled public opposition in Europe against *Operation Iraqi Freedom*. The proliferation of information technology will increase the influence of opinions from states and non-state actors, as well as its use as a means of disinformation.[24] Furthermore, globalization and information technology have facilitated the development of transnational movements lacking central leadership around various issues. Examples of such movements include the global Islamic insurgency, feminist, health and environmental movements. These movements may become accepted as part of popular culture and have wide-reaching influence on governmental policies.

Regionalism

Regionalism and transnationalism are separate but related phenomena. Both are motivated by prosperity and stability. Both are enabled by and dependent upon growing access to information and information technologies. But either can exist without the other and continue despite the other. Regionalism, a growing trend, is the phenomenon that enables a geographically defined region to act as a single entity in order to achieve a common objective. Globalization and the information age are enablers of regionalism. As individual nations see a need to interact with the global community, they question their ability to do so effectively. Often they will form associations with other nations in their region and conduct business as a cooperative grouping. Information age technologies facilitate the formation, coordination, and collective action of these entities.

In the current global environment, economic growth and prosperity appear to be the most common motivation for regionalism. Collective security against external threats fosters a regional approach, but only as long as that threat exists. A common concern over a shared regional resource, such as water or the environment in general, can result in regionalism focused internally. In any of these cases, cooperation in one area tends to encourage cooperation in the others. Regional groupings gain strength when based on several interrelated concerns (for example, security and prosperity) and continue to strengthen over time.

Terrorism

Physical, psychological, informational, and economic terrorist targets will threaten state governance in some regions and the form, organization, and tactics will evolve beyond that experienced today. Often, many groups who previously were unable or unwilling to use terrorism may resort to it as the only mechanism to further their cause and combat external influence or threats. Although some terrorist organizations may receive support from states, (e.g., Iran) terrorist organizations normally guided by ideologies with a regional or global message and directed by core groups (the jihadist enterprise, guided by al Qaeda's ideology and leadership are current examples) will remain the dominant terror threat for the

---

[24] Anthony Barnett, World Opinion: The New Superpower? Open Democracy Ltd, Mar 2003

foreseeable future. The current insurgency in Iraq is creating fungible insurgent skills that will eventually disperse terrorism throughout the world through its veterans and via the jihadists' online distance-learning enterprise."[25]

Temporary security alliances along the lines of a "coalition of the willing," may form to respond to regional or international terrorism. Although the U.S. will remain the dominant military force, lethal niche capabilities could allow small states and non-state actors (including terrorist groups) to form temporary alliances or coalitions based on common ideology or objective that will threaten the deployment and mission accomplishment of U.S. armed forces.

Alliances and Coalitions

Future alliances and multilateral arrangements will involve a greater array of actors than at present. For example, the Shanghai Cooperation Organization (SCO) brings together major Asian powers and central Asian states into a coalition dedicated to minimizing the influence of outside powers in the region. Uniting or disbanding based on common interests, new forms of alliances and coalitions will be a common feature of the future international environment, and will use collaborative information sharing and database development as a primary means to unite in purpose. Traditional state sovereignty will play a decreasing role. International organizations, regional supernational organizations, nation-states, NGOs, local leaders, MNCs, special-interest groups, and religious organizations may be part of future coalitions, friendly or adversarial. Members of alliances and coalitions will seek to control and focus the actions of the U.S. It is also likely that U.S. armed forces will be tasked to reconstitute governance or substitute for established governance.

2.3 Resources and Economics

Climatic Disruption

Climate change driven by global warming will have wide-ranging economic and resource impacts. By one estimate, increased extreme weather could reduce global gross domestic product (GDP) by up to 1%, while a worst case scenario could cut economic growth by 20% The IPCC projects macro-economic costs associated with mitigation of greenhouse gases by 2030 as between a 3% decrease in global GDP and a small increase. The economic impacts will range from those associated with resource availability, to increased health care costs, and the potential failure of the insurance industry, the world's largest economic sector.[26]

Since the 1970s, insurance losses have increased at about 10 percent each year, with destructive weather, including heat waves, hurricanes, typhoons, tornados, floods, wildfires, hailstorms and drought accounting for 88% of all property losses paid by insurers from 1980 through 2005.[27] The frequency and severity of all these types of events is projected to increase due to climate change. The world's largest reinsurance company, Munich Re, has estimated that by 2050 the global damage bill from climate change could top $500 billion. Insurance coverage provides stability for governments and businesses as well as individuals.

---

[25] Jenkins, Bryan, "Terrorism: What's Coming The Mutating Threat, Memorial Institute for the Prevention of Terrorism, 2007
[26] Stern, Nicholas, The Stern Review on the Economics of Climate Change, 30 October 2006
[27] Flannery, Tim, "The Weather Makers", Atlantic Monthly Press, 2005

As insurance companies reassess risk and abandon risk prone markets, the resulting cost reallocation will produce widespread economic as well as political turmoil.[28]

Figure 2.5. Intensity of Global Climate Change

Adequate supplies of fresh water for drinking, irrigation, and sanitation are the most basic prerequisite for human habitation. Many regions are already suffering water shortages due to drought, agricultural diversion, pollution and other reasons. 40 percent of the projected world population will live in water-stressed countries by 2015. Climate-induced changes in rainfall, snowfall, snowmelt, and glacial melt have significant effects on fresh water supplies, and climate change will affect all of these areas. Glacier, runoff, in particular, provides half of the drinking water for 40 percent of the world's population. These glaciers have been receding for some time and increased warming will increase the rate of melting, increasing existing pressures for available water in several regions, notably the Middle East and Asia. By 2030, up to two thirds of the world's population may face water shortages.[29]

Climate change will have multiple effects upon food production. Crop ecologists estimate that for every 1.8°F rise in temperature above historical norms, grain production will drop 10 percent. Food production, cultivation and animal husbandry patterns will be affected and some regions will be unable to grow current food staples, such as rice and green vegetables; fish stocks will diminish or migrate.[30]

Economics and the natural resources required to sustain economic activity and standards of living around the world are central to an understanding of instability, war, and the will and ability of states, organizations, and individuals to be involved in conflict. The current world economy is characterized by the notion of *globalization*, which denotes the ability to trade, conduct commerce, and move goods and services across international boundaries. Globalization has brought with it a degree of prosperity that has never been seen before. It also brings with it economic dislocation as centers of high-cost production are closed and moved to areas with lower labor and production costs. Because globalization results in both "winners" and "losers," the degree to which economic globalization will continue is unclear.

---

[28] Morrison, John, and Alex Sink, The Climate Change Peril That Insurers See, Washington Post, 27 September 2007

[29] National Security and the Threat of Climate Change, CNA Corporation, 2007

[30] United Kingdom, The DCDC Global Strategic Trends Programme 2007-2036, The Development, Concepts and Doctrine Centre, 2007

Success depends on sustained development and a method to ensure that "losers" are effectively transferred to new and profitable economic activities. It also depends on the expansion of accepted rules of international contract, property, and civil law and the equitable distribution of economic gains.

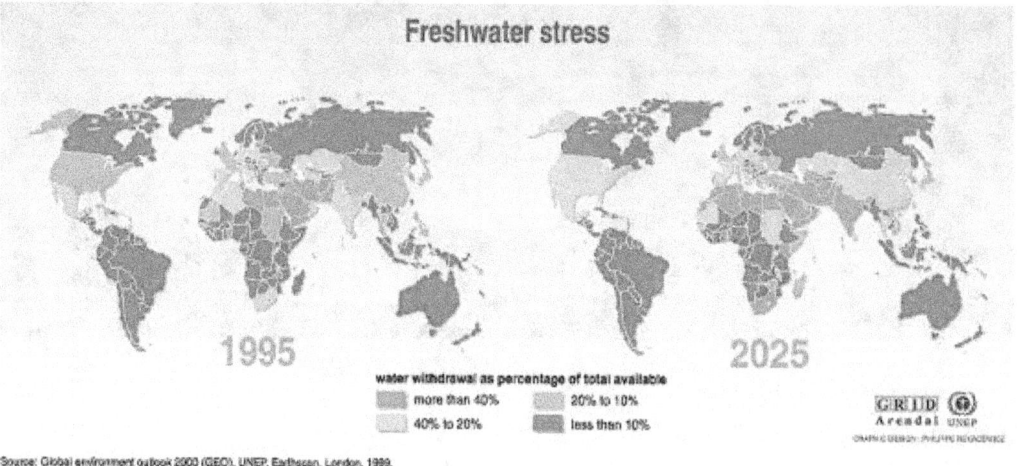

Figure 2.6. Stress on Sources of Fresh Water

The following trends in economics and resources will influence the likelihood and character of conflict and war in the future operational environment. The joint forces will be required to consider the impact of these trends in its planning and operations.

Resource Competition

During the next 25 years, there will be a shift in the pattern of resource dependencies. In the developed world, political and environmental concerns, in concert with technological improvements, will lead us to seek a reduced reliance on fossil fuel. However, even by 2030 hydropower, non-ethanol renewables, and nuclear generation will account for 1/5 of total energy use. The greatest rise will occur in ethanol use for vehicles, which will increase to 25 percent of total vehicular energy use from today's four percent.[31] Even though many of the energy efficiencies available to the first world will also be available to developing countries, their increasing needs will lead to a greater demand for oil. India and China will see their energy demands rise to "first-world" levels. As these developing nations prosper, energy demands will grow as a result of increased heating, cooling, industrial, and transportation needs. Natural gas, coal, and oil consumption have all grown worldwide. Potential chokepoints exist in transportation of those fuels between producer and consumer nations. As consumption of these fuels has increased so have carbon dioxide emissions. Besides the fact that these resources are finite, the developed-developing nations must face the impact development has on a shared environment. There will be increased competition or perhaps conflict over energy resources, which may drive military and political priorities. The criticality of energy resources and their attractiveness to some actors as targets of disruption will lead to armed conflict both in the resource areas and in and among distribution systems.

---

[31] U.S. Energy Outlook 2008. *Annual Energy Outlook 2008*. (December 2007).

The economies of several large oil exporting countries are growing so quickly that by the middle of the next decade, many will use all internal production for their own economies, leaving none for export. This explosive growth means that several of the world's largest suppliers will begin importing oil to power their economies, placing further strains on the global oil economy. One country that has already made the transition from major exporter to net importer is Indonesia. The same "flip" may likewise occur in Mexico over the next decade, which is currently the second largest exporter of oil to the United States. Rising internal demand may take 40% of Saudi Arabia's increased production.[32]

Issues of resource management (water and energy sources) become significant in regions when population demands outstrip local resources; e.g., Darfur, where wells are drying up and most wood has been burned as fuel. Technology, alternative energy sources, and improved conservation methods will provide some relief, but potential conflicts over scarce resources could easily destabilize some regions. Access to resources will continue to be a primary concern of every state, and competition for limited resources will be a cause of future conflict. States without large supplies of energy, water, mineral wealth, and agricultural resources will be significantly challenged to maintain economic growth and prosperity. Some adversaries might be able to leverage such situations by attempting to create instability in those countries controlling or using resources.

Shortages of food and water will create problems in many regions and countries. Currently a billion people lack access to fresh water. About a dozen nations in Africa and Asia will experience severe water scarcity by 2025, fueling mass migrations, humanitarian crises and conflict. Large amounts of arable land are being lost to desertification, erosion, salt intrusion, and urban sprawl, which could limit crop yields. Sixteen percent of world protein comes from fisheries, with considerably more in some developing nations and in regions that depend heavily on the sea, but this resource is under stress from over fishing, pollution and other environmental factors.

As economic systems become more integrated, interdependent, and globalized, they will be increasingly vulnerable to intentional disruptions to the supply of vital resources. Developed countries, no longer producing much of their own electronics, steel, and energy will be particularly vulnerable to interruption of the movement of vital materials at critical points. Control of pipeline pumping stations, maritime chokepoints, major ports, airfields, key rail junctures, and other critical segments of the transportation infrastructure will be vital, especially during a crisis. Internal conflicts may also arise over resource distribution and management, especially because of inadequate distribution infrastructure. Ironically, countries that are or will be increasingly prosperous could be increasingly unstable because of inequitable distribution of their new wealth; such instability will likely lead to discontent, rebellion, and migration.

Distribution of Wealth

The 2005 UN Human Development Report creates a picture of growing inequality and an increasing global gap between rich and poor. The richest 50 individuals in the world have a combined income greater than that of the poorest 416 million The 2.5 billion people living

[32] New York Times (December 9 2007)

on less than $2 a day – 40% of the world's population – receive only 5% of global income, while 54% of global income goes to the richest 10% of the world's population. But the problem is not just one of inequality between countries. The HDR points out that in the last 20 years the unequal distribution of income within many countries has grown worse. Of the 73 countries for which figures are available, 53 (comprising over 80% of the world's population) have recorded an increase in inequality of distribution. Despite globalization of networks, there is no guarantee of effective or even distribution of wealth. For example, only in 9 countries (comprising about 4% of the world's population) has the wealth gap between rich and poor been at all reduced. Differences are especially great within Namibia, Brazil, South Africa, Chile and Zimbabwe. Even in countries with high economic growth rates – Brazil and China for example - social disparities remain large. This rise in wealth disparity will lead to increased tension and hostility between the rich and poor in the future, as the poor seek relief from any available source[33] Furthermore, the global information environment means that the poor have an ever-greater awareness of their plight and can readily compare their circumstances with the world around them – further causing disenchantment, frustration, and possible conflict.

Global Trade and Financial Links

Global commerce may be characterized as three overlapping network layers: physical, financial, and informational. The networks are efficient and connected, but are increasingly fragile and difficult to rewire. The networks demonstrate characteristics of a scale-free network consisting of bridge nodes to clusters. Such networks are robust against random failures, as networks can reroute themselves or be rerouted; however, they are susceptible to directed attacks against critical nodes, such as a major container port. Network disruptions do not need to be large to have an impact. Lack of space and capacity for recovery magnifies the disruption; the effects of the disruption will persist and propagate; and multiple small disruptions can cause cascading effects that can reinforce each other, leading to system failure. In the case of the global economy, such network disruption could have wide-ranging effects.

This increased interconnectivity in world markets, which consequently affects local decisions and policies, may have unintended global consequences. The increasing complexity and speed of the global economic system and limited resources and markets will give more significance to economic and resource alliances and blocs. Competing for access to markets and influence will increasingly become a joint governance-business issue, while traditional national and international economic mechanisms will be less effective. Change will be better managed in the developed countries, but in some developing countries the situation may become so complex and volatile that traditional economic assistance will not work without major political and social change—and possibly military intervention.

Emerging markets could cause a range of potential instabilities and shocks. A generation ago, the health of the global economy depended largely on the financial stability of a small number of powerful democracies. Over the next generation, it will depend on political stability in a growing number of countries that have little in common beyond unconstrained growth and the potential for domestic turmoil. Instability in emerging markets will occur

---

[33] United Nations Development Program (UNDP). (2205)*Human Development Report: Deepening democracy in a fragmented world.*

within the context of current high market liquidity, in which nothing looks risky. When this changes, it will lead to contagion in other markets. Change may be a surprise in that it may be rooted in fast-moving market rumors that can become self-fulfilling, which could lead to cascading effects from things that have not even been considered.[34] Today, more capital flows around the world than at any time in the past century. Since 1973, cross-border capital flows have grown from 5% of the global economy to 21%.[35] For the U.S., what was in 1989 a stream of capital moving in and out of the domestic economy has become a torrent. The value of foreign stocks, bonds, and factories owned by Americans at the end of 2006 reached 13.7 trillion, up from just 2.1 trillion in 1989.[36]

Tensions over monetary, fiscal, environmental, trade, and safety and security issues will exist among national governments, businesses, and international organizations. Aggressive capitalism, globalization, blatant consumerism, environmental issues, and public health will exacerbate tensions and result in a blending of roles and responsibilities. Business will acquire greater leverage with governments and international organizations, because it is more flexible and has more options. Corporations may form coalitions to support or oppose governance. An example of this is are oil companies that partner with a local national military, and pay the salaries and expenses of a special armed and uniformed national police force tasked with guarding oil industry facilities. These are not company security guards but national security forces answerable to the government. Nationally important businesses may assume a greater role in national security decision making. Economic actions in one locale may spark conflict in a distant locale. The security forces of large, multi-national corporations will be definite considerations for the use of military power.

In addition, the day will come when Multinational Corporations (MNCs) will purchase commercial intelligence and sell or employ surrogate or mercenary forces to exert influence and to wage conflict. Information technology will enable corporations, governments, or groups to coalesce quickly to form political and economic blocs in response to change. Large, temporary "single-issue" coalitions will be able to communicate and organize at both the macro and micro levels. Unilateral actions by outside players may be more readily blocked or diverted, adding significantly to the number and type of actors that will be able to influence military operations.

Information-Age Economics

Change will become more rapid and often discontinuous within a complex, interconnected, global, technological environment. Countries and organizations will need to be flexible to manage change or they will fail. Command economies that are planned and controlled by a central administration (such as North Korea) or traditional societies (such as Yemen) will be hard pressed to keep pace. Trade volume will increase, as well as the number of players and their impact on world trade. The stock and commodity markets will become more vulnerable to short-term manipulation as IT permeates the global environment. During times of crisis or war, adversaries will use market manipulation to support or frustrate international actions or merely to garner quick returns. Adversaries will strive to profit from

---

[34] U.S. Joint Forces Command, Summer JOE 2007, Exploring Economic Trends and their Implications for the Future Operational Environment.
[35] Federal Reserve Governor Frederic Mishkin, *The Next Great Globalization*.
[36] USA Today, December 10, 2007.

or affect markets as a way to improve their economic position while reducing that of the U.S.

Technology will have both economic and military impact. New technologies can be the engine for rapid economic growth, but they have a price. Increasing amounts of capital must be spent merely to keep up with technological change. Money invested in legacy or inappropriate technology can retard a nation's ability to respond to rapid change. Militaries tend to have long-term R&D cycles, while business works on shorter-term goals. Potential foes with access to business will benefit.

Economic Regionalism

Regionalism contributes to the growth of the global economy. It allows individually insignificant nations to cooperate and deal as a single entity on the global stage. This fosters economic growth throughout the region and across the globe. Rapidly growing access to affordable information technology and information encourages regionalism. It enables nations to discover and discuss common issues and develop a coordinated approach to those issues. Globalization will tend to accelerate the trend of regionalism. As the world economy becomes increasingly interdependent, the global marketplace becomes both more accessible and more competitive. The small, independent player risks becoming marginalized. Smaller nations will recognize that they cannot act alone. They will tend to form economic relationships within their region that allow the region to deal collectively with allies and competitors.

Regionalism requires and promotes stability. Often, when confronted with internal unrest or external threats, a nation or group of nations cannot adequately focus resources on solving economic or social issues. When regionalism takes hold, it fosters stability. The more facets of regionalism bind together a region, the more those bonds will work to encourage continued stability. Regionalism mitigates conflict. Focused on internal cooperation, regions can deal with a wide range of ethnic, resource-sharing, demographic, social, and environmental issues that historically lead to competition and potential conflict.

Global Labor Markets

Labor markets will be in transition. The transfer of industry to developing countries can bring prosperity, or the promise of it. As previously noted, pools of unskilled and skilled labor will compete for jobs, and workers will migrate to affluent countries while jobs migrate to poor countries. Corporations that provide benefits and job security may command more loyalty than governments. Industry will continue to move among work forces based on cost effectiveness and the ease of relocation. Technology will facilitate this movement, allowing less-educated and less-skilled workers in underdeveloped regions to perform similarly to skilled workers in developed locations. Some regions, such as Europe, have an aging skilled workforce, while other regions, such as South Asia, have a young unskilled workforce with few employment opportunities. Increasingly, the constant migration of labor among countries and organizations will enable adversaries to implant and use, when appropriate, sleepers, intelligence collectors, deceivers, cut-out operatives, and direct-action personnel.

2.4 Science, Technology, and Engineering

The key strategic implications of science, technology and engineering (ST&E) will be shaped by and dependent on its global availability. Advances in science and technology will accelerate globalization, thus further increasing the divide between global winners and losers. Technological progress will allow weapons to be more easily concealed and delivered. Culture will have a great influence on ST&E and vice versa as different cultures develop, adopt, and exploit technology in different ways. Economics and other factors will influence ST&E as well because many key developments and breakthroughs will occur commercially. For example, worldwide government R&D spending is dropping significantly.[37] Increasingly, offshore manufacturing will outstrip U.S. domestic manufacturing. The next "key" technology is unknown, but there is no guarantee that it will be discovered or exploited in the U.S. Thus, military scientists and researchers will have to have pervasive, sustained, and trusting relationships with the commercial sector, at home and abroad.

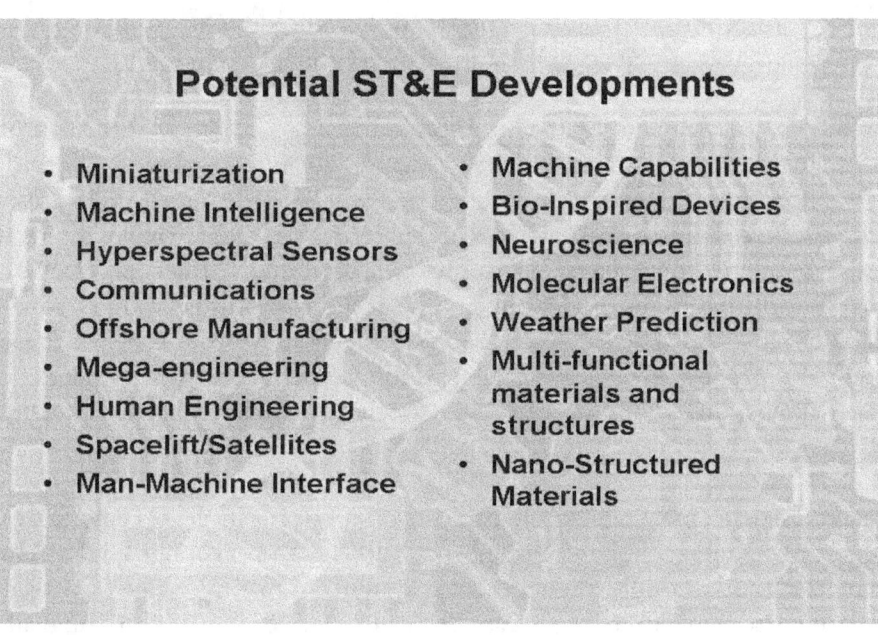

Potential ST&E Developments

- Miniaturization
- Machine Intelligence
- Hyperspectral Sensors
- Communications
- Offshore Manufacturing
- Mega-engineering
- Human Engineering
- Spacelift/Satellites
- Man-Machine Interface

- Machine Capabilities
- Bio-Inspired Devices
- Neuroscience
- Molecular Electronics
- Weather Prediction
- Multi-functional materials and structures
- Nano-Structured Materials

Figure 2.7. Science, Technology, and Engineering Fields

Multidisciplinary technologies across dimensions will have a revolutionary impact on how we live by 2025, but will accentuate the disparity between "haves" and "have-nots." Research requires consistent effort and resources. It cannot be turned on and off effectively or in a timely manner. There is concern that the vagaries of U.S. government funding could put research at risk. Increasingly, many of the best and brightest advanced science and engineering students attending U.S. universities are foreigners, with the number of engineering full-time graduate students without U.S. citizenship actually exceeding U.S. engineering graduate students.[38] Most of these students return home at the completion of

---

[37] 2005 Global R&D Report. *R&D Magazine Online Edition* page G4.

[38] U.S. Department of Education/National Center for Education Statistics: Integrated Postsecondary Education Data System Completions Survey. The percentage of nonresident aliens receiving master's degrees in science and engineering in the U.S. was 26.8 percent in 2000 compared to 23.4 percent in 1991; National Science Foundation/Division of Science Resource Statistics. Graduate Students and Postdoctorates in Science and Engineering: Fall 2001. The percentage distribution for doctoral degrees in science and engineering for

their studies. The U.S. still leads other countries in R&D but 70 percent of world R&D is conducted outside the U.S., and the share of U.S. R&D in the world is on a downward trend.[39] While not necessarily portending a brain-drain in U.S. know-how, off-shore capabilities will increase, often approaching those of the U.S. For example, according to the American Academy for the Advancement of Science, China has increased its R&D investments dramatically in recent years and is now the third largest investor in research and development (adjusted for purchasing power), behind only Japan and the U.S.

The pace of global revolution in science, technology, and engineering (ST&E) development is expected to accelerate during the next two decades. This rapid rate of change will remain the hallmark of ST&E for the foreseeable future as innovative discovery continues within all scientific fields. The ST&E world can be divided into six major subject areas: energy biological systems; machines and computers; information, knowledge, and communications; weapons of mass effect; and environmental science. These discussion areas are broad, but in many cases, the connections or fault lines between them hold the most interesting insights for a military force that is anticipating the future operational environment.

Advances in ST&E will provide significant improvements to many aspects of future life. Things will be smaller, lighter, smarter, faster, cheaper, stronger, and more efficient than they are today. Multi-functional materials and structures will become increasingly important. Nano-structured, or very small-scale structured, materials will help drive many of these developments. Similarly, the scale of electronics will be altered. Molecular electronics, using molecules to perform functions of electric circuits, is a direction that will improve computational capabilities. The small size and potentially easy production of these extremely small materials and processes will lead to faster and cheaper tools. While there is a trend toward smaller things, there continues simultaneously to be activity at the opposite end of the scale.

Mega-engineering—extremely large-scale projects—will occur in those regions of the world where enough capital and capability can be brought to bear on problems. The Three Gorges Dam in China is a good example of such a large system. While this scale of project is likely to continue, it may become limited to areas where alternative solutions are unavailable. For example, the combination of water conservation and alternative sources of water (perhaps desalinization), combined with the ability to make use of seawater (perhaps for irrigating genetically altered plants), will lessen the need for larger water projects. This could have direct military impact because the potential for a conflict stemming from a competition for scarce resources would likely be reduced. In other areas, swarms of extremely small machines may do the work currently done on a very large-scale, such as mining or remote sensing.

What follows is a brief survey of trends in several key technology areas that will have considerable implications for the future joint force commander.

---

nonresident aliens was even higher at 29.4 percent in 2000. 45,014 Full-time engineering grad students without U.S. citizenship were enrolled in the U.S. during 2001, compared to 32,558 with U.S. citizenship.
[39] Organization for Economic Co-operation and Development (OECD). *Main Science and Technology Indicators, 2002. Shares of Total World R&D, 2000.* World equals OECD members plus Argentina, China, Romania, Israel, Russian Federation, Singapore, Slovenia, Taiwan.

Energy

Energy may well represent the United States' greatest vulnerability. The combination of high petroleum usage, location of known petroleum reserves in unstable regions, and Al Qaeda and Iran's stated strategies require us to take a close look at this issue. Barring a major breakthrough "alternative future" in unforeseen alternative energy sources, the world economy will remain heavily dependent on oil through 2025 as a minimum. The fact that the major supplies of oil will not keep pace with world demand, and lie in unstable and violent areas means U.S. forces must be prepared to maintain and if necessary restore security in key areas of the world to insure the continued supply of oil for ourselves and our primary economic partners. It also means improving our collective energy efficiency and the viability of alternative energy sources will be a major strategic imperative for U.S. services and the American public.

Alternative energy sources will likely become more prevalent than today, but will not replace hydrocarbon energy sources, and although more expensive than oil, can have significant military utility in a variety of circumstances. Hydrogen, various forms of atomic energy, solar cells, and hybrid gas/electric systems could potentially somewhat lower our reliance on fossil fuels. As a result, long military logistic tails would shorten. Combat forces would be able to operate for extended periods without being totally reliant on support units. Sensors and systems will have longer ranges and greater persistence, powered by derivatives of alternate energy sources that replace or enhance current battery technology.

Given the fragility of our crude oil supplies over the next twenty years, expanding world consumption, and the lack of meaningful large scale alternatives to oil, access to and source protection of oil sources will continue to be a major policy and security focus of our nation and our competitors well into the future.

Biological Systems

Biology will continue to be an important area of analysis within the ST&E world. Biological systems and processes will inspire sensors, manufacturing, and self-modifying diseases, and will genetically modify crops, animals and people. "Bio-informatics" will begin to harness biological processes to continue the rapid growth in information technologies. Human capabilities and knowledge (health, strength, and cognition) will be enhanced and improved in many ways. As medical science advances, it will become increasingly easy to select the sex and characteristics of babies. This may be desirable in some societies, as in China today, and may result in a relative shortage of women.

Human engineering will alter the way people will be able to think and act. Those who benefit from it will live longer, healthier lives with much greater potential to provide meaningful contributions to society. The converse, though, may be devastating: the shallowness of perspective based on over-reliance of technology could result in "brain-drain" and the erosion of intellectual capital. Some entities will use advances in technology to seek enhancements that can eventually lead to "super human" strengths, cognition, and senses, while degrading "undesirable" human traits (such as sympathy, emotion, and love). In general; however, human engineering on a broad scale is likely to be constrained through 2030 by a combination of resource and ethical constraints.

Biological engineering of organisms could be used to mitigate disease, malnutrition, pollution and crime. Some actors will use that biological knowledge and its potential power to do harm. The prospects for "designer" biological (and chemical) warfare agents grow with each advance in the biological sciences. The ability to enhance human performance or alter human behavior will be available to any individual or group with the financial and mental capital to exploit it.[40]

Synthetic biology obviously has enormous positive potential for changing every element of our world – from creating plants that will break down completely into ethanol to tailored/inexpensive medicines to organisms that completely break down industrial waste. Potential is limited only by imagination and funding. Unfortunately, the same is also true of negative potential including the potential for use as a weapon. Given access to the science and technology of bio-weapons on the global information grid, it is conceivable that Bio-terrorists can produce significant amounts of biological agents with nothing more sophisticated than a "kitchen sink" laboratory. For the foreseeable future, the ability to produce agents far outstrips the ability to detect them.

Neuroscience, the study of how the human brain processes and analyzes, will contribute to human cognition and health. There will be a growth in understanding the biochemistry of the cell membrane and how information is received and processed. Understanding how information is organized for use and storage in brains (human and machine CPUs) will also lead to cognitive improvements and perhaps linkage of human and machine.

Machines and Computers

By 2020-2030 machine intelligence and capabilities could surpass human capabilities. Robotics will play an increasing role in business, personal activities, and military affairs. Militarily, robotic swarms will become more prevalent as potential adversaries take advantage of now-nascent thinking and developments in miniaturization. Emerging technologies will continue to support or surpass Moore's Law of computing power (data density will double approximately every 18 months), since the increasing rate of change in technology is a critical future trend. [41] Molecular, biological, optical, and eventually quantum computing will eventually start to replace silicon-based integrated circuits. Quantum cryptography may be available within the next 20 years, thereby allowing unbreakable codes to be developed for mass use. Broadly speaking, in the 2030-plus timeframe, humans will be inextricably linked and in some cases it will be impossible to differentiate between man and machine. This phenomenon will be as real for friend as for foe.

Information, Knowledge, and Communications

Pervasive information (information that is available at any time and place) combined with lower costs for many advanced technologies, will result in individuals and small groups having the ability to become "super-empowered." They will employ niche technology

---

[40] Deputy Chief of Staff for Intelligence, US Army Training and Doctrine Command. *Mad Scientist 2004 (TRADOC DCSINT 6th Annual Future Technology Seminar) Findings White Paper* published 05 January 2005
[41] Mike Martin, "Nanowire Circuits Could Spur Computing Advances," *NewsFactor Network*, March 28, 2003. According to Harvard chemistry professor Charles Lieber, use of nanowire circuitry in place of standard integrated circuits will demonstrate potential for extreme performance in electronics well beyond the end of Moore's Law.

(WME, for example) capable of defeating key systems and providing inexpensive countermeasures to costly systems. These super-empowered people or groups will have a magnified ability to do both good and evil. There will be a greater probability that true democracy can flourish in areas that make the best use of available technological opportunities. Concurrently, some individuals or groups will have the ability to exert greater influence than others. Time and distance constraints will become largely insignificant. Super-empowered groups will be able to plan, execute, receive feedback, and modify their actions, all with maximum synchronization.

Communications links, enabled by wireless and broadband technology and connected through vast and complex networks, will continue to grow. These integrated, interdependent systems will provide much of the expanded level of available knowledge. Indeed, information webs will create greater combined intellectual power. The collaborative information environment (CIE) enhanced "collective brain" will come together and disband based on need. Hundreds of minds enhanced by technology and working as one will far out-distance individual geniuses. These connections, more often electronic or remote, but occurring also at a personal level, will improve how people relate to each other. While cultural biases will remain, there likely will be much more and effective communication. For example, automated language translators will become the norm; first for written electronic communications, and shortly thereafter, direct oral translations between people. Modern communications will change the economic picture down to the village level; receivers are now broadcasters and information reporting will be individualized.

The complexity of information systems has a continuum of risks and strengths, however. More webs create greater combined intellectual power, but they also create more interdependencies and therefore more vulnerability. Indeed, as the U.S. military transitions to and becomes dependent on network-centric operations, the complexity of future networks and interactive systems of systems will bring out inherent risks associated with the loss or compromise of information on the network. Such loss could occur through system failure, such as physical disruption of a key node or human error.

Information on the network could be modified (possibly with malicious code) or sensors and processors could be overwhelmed with data input. Presenting more targets in a short time than could be countered would also pose a significant threat. The continued commercial outsourcing of computer code writing for use in military systems will enhance this potential. When a network is stressed in one area, there is greater potential for widespread cascading effects, not always in expected areas. Information reliability becomes crucial to an organization that is dependent on it for survival.

The interconnectedness of the world and the empowerment of certain individuals and groups will lead to a desire by some to influence events and a growing belief that people can affect anything. This belief, accentuated by collective intellect and man/machine symbiosis, could lead to a new, virulent strain of uncontrolled aggressive intellectual behavior, indeed, competition. This is the psychological nexus that technology has with the human mind. At times, and perhaps at all times, this will be seen in warfare. Boundaries of what is acceptable in warfare will continue to blur. Adversaries will seek vulnerabilities in information systems; some unrelated to military use, and exploit them with devastating results. Swarms of micro-size, networked machines may be used to perform intelligence, surveillance and reconnaissance (ISR) operations, and may be used for physical destruction or disabling of an opponent's equipment. Swarming technologies, however, are not limited to the very-small, nano-level. Mini- or micro-sized tunneling underground vehicles, for instance, could swiftly engage underground targets either in direct action or in data collection. Small, low-observable, long-range UAVs will be developed that can be launched en masse with little vulnerability to detection. Microrockets will swarm space-based systems. Even long-range unmanned underwater vehicles loom on the horizon that can be swarmed against our ports and cities. Self-healing, networked minefields will propagate in various terrain sets, impeding our ability to move through the environment.

Models and simulations will be pervasive and provide far more accurate portrayals of reality than previously achieved. Games in synthetic environments with avatars acting as adversaries will become more important for training, education, and interactive wargaming. Models and simulations will be increasingly sensitive to initial conditions and details embedded within them. Networks and network centricity will be fundamental to future conflict. Networks will increase in complexity, pervasiveness, effectiveness, and density/layering, for example, expanding and contracting push and pull methodologies to and from communities of interest (COI) and communities of practice (COP). The use of "electronic data" will increase, and that same data will be vulnerable to attack, destruction, manipulation, or alteration/corruption.

Weapons of Mass Effect

For the foreseeable future, it appears that the existing triad of nuclear, biological, and chemical weapons will remain the capabilities of choice with regard to weapons of mass destruction and mass effect (WMD/E). Of the three, nuclear appears to be the most effective in terms of its ability to kill and destroy infrastructure with one weapon. Biological weapons may be considered the most challenging, in that they are cheaply and easily produced, easily transported and dispersed, both physically and psychologically effective, and their early detection is problematic. Chemical may become the most common and readily available WMD/E, as many industrial chemicals can have toxic effects if misused, intentionally or not.

The future of WMD/E can be seen not only in the advanced technology of weapons development, but also in the application of dual-use technologies and innovative use of existing and emerging technologies to produce WMD/E-like effects. Anticipated and potential advanced technologies include:

- **Bio-engineered weapons.** Future bio agents will be more virulent. Bio-engineering will allow bio weapons to be tailored toward specific targets and groups of targets. The globalization of the world market offers numerous vectors for the effective spread of bio agents, which can have an immediate impact or be tailored to remain dormant for a specified time. The likelihood of bio weapons targeted against materials – oil, rubber, metals - must also be considered.
- **Chemical Agents.** Development of chemical agents continues, with the aim to penetrate/defeat our protective gear, which is tailored to protect against gas and liquid agents.

- **Nuclear Weapons.** 1st and 2nd generation technologies are readily available and required materials are abundant. Research continues to increase electromagnetic pulse (EMP) effects, improve efficiency of the conversion to energy, explore alternative fuel sources, and investigate the potential of anti-matter as an energy source and/or weapon.

- **Directed Energy.** The next 25 years will see widespread operationalization of directed energy weapons (DEW). DEW will be developed to target both people and materials and will include non-lethal lasers and ionized radiation projectors, anti-personnel and sensor lasers, as well as radio frequency weapons targeting Joint electronics.

- **Nanotechnology.** While nano has the potential for development of weapons in its own right, the use of nano technology in combination with chemical, biological, nuclear, and radiological capabilities is perhaps its greatest threat. Nano can facilitate production, concealment, delivery and activation of these more conventional threats, thereby enhancing their effectiveness as WMD/E. . For example, encapsulating biologically altered material in a carbon nanotube for storage and or delivery is possible by 2025. Nanotechnology will also produce micro-cameras, sensors (including chemical, biological, radiological, and nuclear) and communications networks.

- **Geophysical weapons.** Much of our environment can be exploited to produce mass effects equivalent to a conventional WMD/E. There will be technological advances that can stimulate and /or enhance natural disasters such as earthquakes and degrade the environment by deliberately accelerating global warming. Manmade infrastructure – large dams, bridges, etc. – can be attacked to mimic the effects of a natural disaster.

- **Cyber weapons**. The rapid growth and associated dependence on information systems create vulnerabilities to weapons is this domain. Adversaries will deny the use of communications networks, will modify information in systems to reinforce and change perceptions of the user community, and will have the ability to conduct these attacks in ways that can be plausibly denied. Specific techniques that continue to evolve are:
  - *Worms*. Self-propagating malicious code that can automatically distribute itself from one computer to another through network connections.
  - *Viruses*. Code written with the express intention of replicating itself. A virus attempts to spread from computer to computer by attaching itself to a host program.
  - *Trojan Horses*. Hidden functionality that is dormant until turned on, either explicitly (e.g. by outside stimulus) or implicitly (e.g. by a time-out not countered from outside). Often carried out via software that purports to be useful and benign, but which actually performs some destructive purpose when run.
  - *Botnets*. A network of computers that have been "enslaved" by a networking virus to perform malicious tasks as part of a larger, directed effort. This represents the ultimate counter-network network and provides unique and devious ways to disrupt information networks at the system level.

- **Control of the human brain.** Technologies will be developed that can work directly on the human brain, causing a wide variety of effects ranging from benign behavior modification to mind control to immediate death.

Environmental Science

Earth sciences will be capable of much better understanding of all environments, including land, sea, air, and space. More accurate prediction of weather and geologic phenomena will aid in planning everything from physical structures to personal time. However, the increasing density of the world's population as well as urbanization and development of economic infrastructure will increase the impact of natural disasters such as earthquakes, hurricanes, floods, fires and tsunamis.[42]

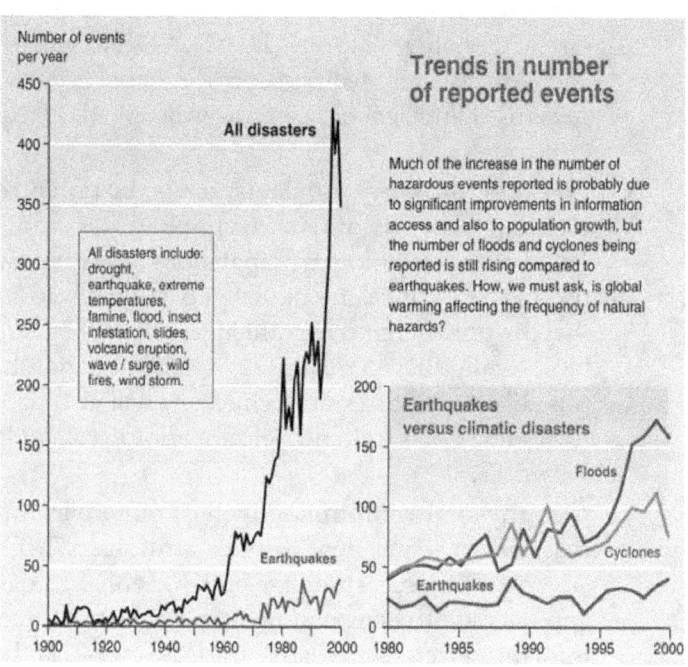

Figure 2.8. More Humans, More Disasters

---

[42] Trends in Natural Disasters. (2005). In *UNEP/GRID-Arendal Maps and Graphics Library*. Retrieved 20:56, December 12, 2007 from http://maps.grida.no/go/graphic/trends-in-natural-disasters.

Chapter 3: Challenges Facing the Future Joint Force

As the most powerful state in the international system and an advocate of democratic principles, free markets, and human rights, the United States cannot expect that all trends under consideration will be moving in a direction congruous to its interests. Various actors with competing interests and priorities will ensure that areas of conflict will arise and where the U.S. will often face serious threats to its security and interests around the world. While predicting the precise outlines of a future conflict can be difficult, and differing conclusions can be derived from similar data, we will explore how combinations of trends will result in challenges to our national security. These future joint force challenges will likely involve the use of U.S. military forces to shape the environment, to deter adversaries, and to apply violence in the service of national interest when called upon by civilian leaders.

The following section provides an overview of the types of future joint force challenges that the country will likely face over the next two to three decades, and are based on the trends previously described in chapter two of this document. The future joint force challenges are placed into three groups as follows:

- Enduring Challenges.
- Emerging Challenges.
- National Security Shocks.

Each future joint force challenge will describe the military problem that results from the confluence of a number of trends and illustrates a set of opportunities that might result from such an environment.

3.1 Enduring Challenges

This group refers to military challenges which are currently ongoing, "obvious," and result from an international system based on the relative primacy of the state. This group of challenges includes historic or "core" missions that the U.S. military is traditionally called upon to solve for the Nation – and will likely remain enduring features of the security environment over the next twenty to thirty years.

Attacks on U.S. Territory

The first and foremost challenge to the joint force is defending the United States itself. The United States has a long history of understanding homeland defense as beginning well beyond the borders of the republic. The Monroe Doctrine discouraged new colonial acquisitions in the new world while our overseas forward basing posture during the Cold War limited the ability of the Soviet Union to dominate Western Europe and East Asia. Both national strategies supported an American strategic worldview that wars should be fought in Eurasia, rather than in the Americas, and that America was most secure in a world open to global trade. The attacks of 9-11, and the real and faked anthrax attacks that closely followed were one of the few major attacks on the U.S. homeland that impacted the society as a whole.

The globalization of trade, finance, travel, and information have brought the front lines of homeland defense closer to home as North America is no longer shielded by geographic distances and naval power from the military activities of our adversaries. The ability of competing interests to impact the United States will likely focus on three central elements of the U.S. homeland; the will of our political leadership and civilian population to engage in the world; our economy as an underpinning of the will and ability to use military power; and the physical capabilities that underpin our ability to project military power abroad.

Homeland defense will include significant interaction with law enforcement authorities for information sharing and support purposes. Should "super-empowered" Special Forces or terrorists engage in sabotage operations or operations to inflict mass or sustained casualties (such as multiple sniper attacks) military forces may be called upon to provide ISR or other capabilities to local officials. Furthermore, states or other groups may seek to deter and dissuade using ballistic or cruise missile attacks on the homeland at either intercontinental or intermediate-range distances, or may mount advanced missile systems on submarines or even commercial shipping vessels. The internet also provides a direct means to influence or disrupt commercial or military activities within the United States, and may require a national military response.

Opportunities
- "Hardened" homeland security may deter and dissuade attacks against the United States.

Conflict with other Great Powers

The ability to dissuade, deter and ultimately defeat great power opponents is a potential feature of the future operating environment. Although interstate conflict -- especially between great powers -- is an increasingly rare phenomenon, tectonic shifts in the international environment including the rise of China and the potential decline of military and economic elements of American power may give rise to new and powerful state challengers. [43] After the collapse of the Soviet Union, overwhelming U.S. conventional military superiority has been a key feature of the international environment, however the spread of military and civilian technology, growing economic power around the world, and the control of key global resources, such as petrochemicals mean that states such as China, India, and Russia may challenge U.S. dominance of the international system and build traditional military capabilities to assert their interests regionally and globally.

> "The likely emergence of China and India as new major global players—similar to the rise of Germany in the 19th century and the United States in the early 20th century— will transform the geopolitical landscape, with impacts potentially as dramatic as those of the previous two centuries. In the same way that commentators refer to the 1900s as the American Century, the early 21st century may be seen as the

---

[43] Most interstate conflicts involving the United States are with significantly smaller powers. Since 1980, the U.S. has been involved in open interstate conflict with Grenada, Libya, Iran, Panama, Iraq (twice), and Yugoslavia.

time when some in the developing world led by China and India came into their own."[44]

This challenge will feature states with powerful conventional military capabilities that may mirror U.S. capabilities and include information-enabled networked forces, naval forces including air and undersea capabilities. These powers will have the capabilities to reach into space and cyberspace, and may be able to challenge the U.S. for dominance in these areas. Emerging great powers will seek to project power farther from their borders and develop expeditionary capabilities to secure energy sources and supplies of natural resources. Emerging great powers will also rely on niche capabilities or local technologies to press geographic and societal advantages and to defeat perceived U.S. vulnerabilities in a number of areas.

Opportunities
- Primacy of economic factors in great power status encourages states to invest in stable world order. Stable world order may allow the U.S. to encourage a "concert" of great powers to maintain global stability and order.
- Flexibility to conduct offshore balancing against Eurasian Powers.
- Decreasing focus on land power/increasing focus on "global commons" (air, sea, space, cyber).

Collapse of Functioning States

Great powers and international leaders will not be able to escape the implications of failed and failing states. The inability for a number of states to cope with the stresses of the changing international environment and their resultant inability to address the needs of their citizens presents a number of difficulties which may draw in U.S. forces, including the use of these territories as bases for global terrorist groups to train, equip, and plan for attacks against the United States and its interests. Failed and failing states are likely to be under pressure from sub-or trans state actors, including ethnic or religious groups, tribes, criminal elements or other identity groups. State failure can also be a result of the environmental pressures discussed above. Many states that struggle to maintain a strong central authority face severe economic, religious, or cultural pressures to exchange conventional order for another type authority, often bringing about tremendous local and regional instability. This complex environment will place a premium on understanding the dynamics among competing groups and an understanding about how the joint force can manipulate these relationships to its advantage. The goal of these operations will often be to reintegrate these societies into the international community and leave behind a functioning state government that is capable of addressing these problems locally. Of overarching concern are states with current or nascent nuclear, chemical, or biological weapons capability, and those with critical global resources. In these cases, more immediate and direct military action may be essential to regional and global stability.

---

[44] National Intelligence Council, Mapping the Global Future: Report of the National Intelligence Council's 2020 Project (Washington, D.C.: U.S. Government Printing Office, December 2004), p. 47.

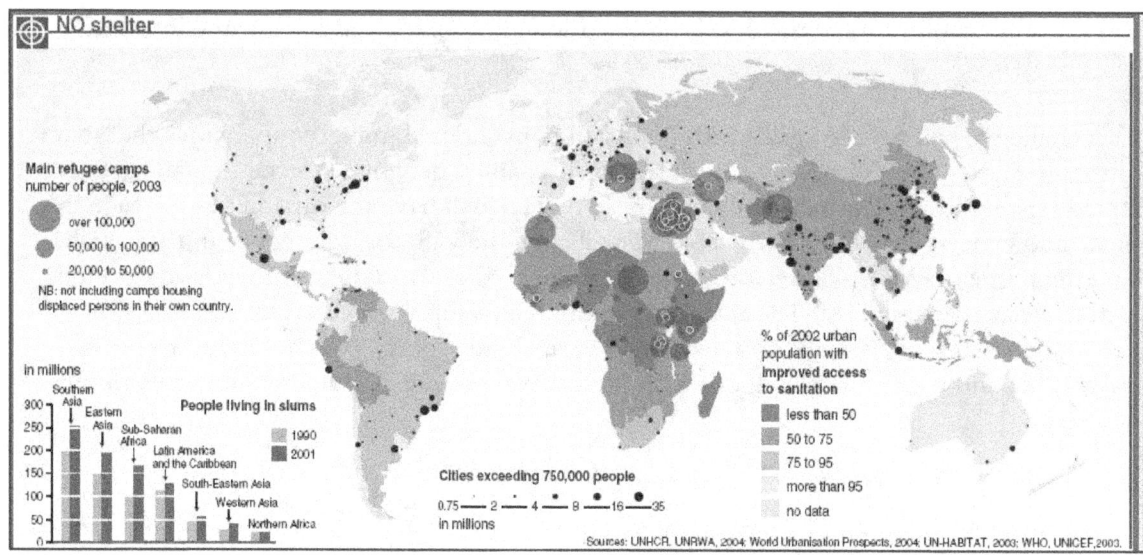

Figure 3.1. Effects of Failed Governance

When the US military operates in a failed state or into areas torn by conflict, it will find non-government organizations (NGOs), international humanitarian organizations, multinational corporations, transnational organizations, and other civilian organizations at work. These external organizations can have both stated and hidden interests and objectives that can either assist or hinder US mission accomplishment. Each organizational or individual participant pursues its interests and objectives in concert or in competition with other entities. Organizations and individual actors may have economic, political, religious, cultural or private motivations, such as revenge, which differ from their public organizational mission statements. Joint force commanders will be required to understand these diverse and competing aims, and understand the impact that these non-governmental "service providers" may have on the accomplishment of military missions.

Opportunities

- Successful ability to provide just and secure governance and development in crisis regions can limit the ability of terrorist networks to operate.
- May result in significant goodwill for the U.S.
- Ability to coordinate with NGOs in this area may allow the U.S. to focus on security and lower costs of repairing failed states.

Conflict with Terrorist Networks

Transnational terrorist networks will remain an acute military challenge for U.S. joint forces. Terrorist networks are already major players in ongoing conflicts, can defeat great powers strategically, and will have the potential to devastate entire populations because of the diffusion of technology and weapons of mass effect. Wide-ranging ideological groups have discovered how to form global networks that operate beyond state control and have acquired the tools and means to challenge states in a number of ways. The current international system is not equipped to deal with the non-state challenge, which is based on the premise that negotiations take place between nation states. It is only just beginning to adjust to the concept of armed groups with effective political control within the boundaries of existing, weak states. The weak states simply lack the legitimacy to speak for the armed groups – or the power to enforce any agreements made. Since the international diplomatic system is not set up to deal with these groups, the military will have to. In addition to weak states, transnational terrorists may increasingly locate bases of operations in the noise of large, developed cities such as London, Hamburg, or Los Angeles.

These groups use tactics that include insurgency, suicide bombing, hijacked aircraft, and improvised explosive devices against civilian populations. Furthermore, they use global communications networks, including phone and internet systems to plan, coordinate, and propagandize for new followers. They will direct these capabilities against symbols of U.S. military, financial, economic, and cultural power. Currently, the most violent and ambitious global terrorist network (and one that will likely be with us for some time) is based around a violent interpretation of the Islamic religion and seeks to establish a new caliphate that transcends current state boundaries in the Middle East.

> *The balance of power will change; the international system built-up by the West since the Treaty of Westphalia will collapse; and a new international system will rise under the leadership of a mighty Islamic state.*[45]

These groups rely on failed or failing states as sanctuaries from which to base their operations and work to exclude and dissuade the United States from engaging, trading, or otherwise accessing the area of their future state. The goal of this exclusion is to undermine current states in the area and mobilize their resources to further disrupt the United States and its economy and society. The threat posed by terrorist networks will remain a pervasive characteristic of armed conflict and military operations, and new ideological movements may take up this method of warfare to further their own causes. Ideological groups capable of constructing transnational terrorist networks will:

- Have significant financial assets at their disposal.
- Have easy access to the global economy and can take advantage of technology and the permeability of borders.
- Not be bound by norms, modes, and methods of international law.
- Be active when the state's capability to exercise the exclusive use of force is not significant.
- Be capable of mobilizing public opinions in their own favor against the community of states using transnational ideologies or religions.[46]

The dilemma for the joint force commander is that the transnational terrorist network conduct attacks at many places, often at the time of their choosing, and will employ a wide range of capabilities to do so. Terrorist networks usually take on an indirect approach, and avoid U.S. strengths and attack at vulnerable or even non-military points. While terrorists will seek to strike relatively secure military targets and installations to create fear and raise questions about security in the minds of the civilian populace and political leaders, their focus is usually to attack public support for the United States, and to separate local

---

[45] Al Qaeda ideologist Lewis Atiyyatullah in 2005 after the Madrid Bombings. Quoted in Grahame Thompson, Working Paper No. 14"Religious Fundamentalism, Territories, and 'Globalization.'" *Centre for Research on Socio-Cultural Change* (U.K) (February 2006).
[46] Bundeswehr Transformation Center, *Outlook to 2035: Trends and Developments.* p.10

populations from friendly (and state-centric) allies. It is clear the numerous armed groups that keep appearing will represent a challenge to the way traditional governments carry out business. Worse, the truly alarming variety of armed groups active in the world today dramatically increases the difficulty of understanding their motivations, methods and goals.

Opportunities
- We have to focus on integrating our political, military, and economic power if we are to minimize their impact on our security.

Conflict with Transnational Criminals

Worldwide, drug cartels and gangs are growing in power and capability – and are evolving into more dangerous entities. In the same way that warfare evolves in parallel with the political, economic, social and technical aspects of a society, so does crime. Criminal networks have expanded along with their increasingly transnational networked business counterparts. All of the elements encouraging global connectivity – trade treaties, cheap transportation, nearly free communications, and increasing movement of peoples – also enhance transnational crime. The massive increase in wealth and improved communication between communities have enhanced international security however this environment also provides enormous opportunities to criminals.

International criminal networks—driven by profit and facilitated by private enterprises that operate through complex ownership structures and participate in diffuse and opaque global supply chains—contribute to such threats as the drug trade, money laundering, corruption, and the trade in weapons. Many terrorist groups, substate actors, or even elements of states (such as the AQ Khan nuclear network, or the Iranian Republican Guards Corps) have turned to criminal activities to support their operations. Criminal coalitions are complex networked organizations that the joint force is encountering today in places like Afghanistan and Iraq, and will likely be a challenge to the joint force for the foreseeable future.

The most notorious of all criminal networks are those of drug cartels. These networks play a major role in a large number of the developing nations and defy efforts by weak governments to eliminate them. The fundamental problem is that as long as there is a demand for drugs, someone will provide a supply. Drug cartels can even strive to supersede the state as the primary authority. They have expanded their coercion options by feeding corruption in their regions. They have simply purchased or rented key elements of government – police, judges, etc. They also make use of very selective violence to eliminate anyone not willing to be corrupted. The choice is simple – take money or die. There is legitimate concern that these organizations are creating narco-states in some of the federal states of Mexico, parts of Afghanistan and perhaps Haiti. They make use of their secure bases and extensive transnational connections to produce, package, move and ship their products worldwide. Clearly, the cartels have evolved, and will continue to evolve in response to perceived threats to their business. Their demonstrated adaptability means they are continually reaching out to other criminal elements, insurgents and terrorists to see if they can make profitable arrangements.

Criminal gangs have taken over of entire sections of Latin American cities. In Brazil, 80 of the 600 *favelas* (slums) are now run by criminal organizations and, like its Middle Eastern counterpart Hizb'allah, sometimes provides more social services for the poor than the government. Of concern is the possibility that, like Hizb'allah, they will establish areas where government forces cannot operate. In May 2006, imprisoned Brazilian gang leaders launched coordinated attacks on Sao Paulo police stations and public transportation to protest their movement to solitary confinement. While the government was able to bring the city under control, this gang, the First Command of the Capitol, demonstrated a clear capability to challenge the government in Brazil's biggest city.

Opportunities
- Encourage "whole of government" approaches to security.
- Strengthen ability to leverage differences of interest between adversaries and "mere" criminals a potential vehicle to disrupt terrorist financial and transportation networks.
- Test cases on how to identify critical areas of vulnerability in culturally imbedded pervasive networks, defeat them, and replace these networks with legitimate systems that support governance and the rule of law.

Prevention of Conflict

The ability to shape the international environment short of war will be a key challenge for the Joint Force. Responding to adversary activities and conducting operations in war is only half the challenge for U.S. forces. The other half will be in influencing adversary and friendly states, and transnational terrorist and criminal organizations or individuals by the presence and activities of U.S. forces during "peacetime" or at least in "pre-crisis" periods. Conflict prevention will call on the joint force to build relationships, share information, and build modes of potential operation in the case of contingencies with regional actors, and security cooperation should be integrated in combatant commander's theater engagement plans. These missions might draw on all elements of national power to favorably dispose these actors to working with the United States.

A second level of conflict prevention includes deterrent operations. These operations include a wide range of activities designed to dissuade an adversary from pursuing some course of action. Deterrence can be conducted using a wide variety of means, including foreign naval presence, foreign basing, and demonstrations during exercises, or in nuclear response policies. The growing trend toward nuclear proliferation among states, as well as the potential leakage of nuclear weapons to non-state actors will require more comprehensive deterrent strategies.

Opportunities
- Encourage more regularity and less violence in the international system.
- Cost of peace is cheap compared to the significantly higher cost of war.

3.2 Emerging Challenges

This group refers to a set of rising challenges that are the result of globalization, uncertainty, complexity, interconnectedness, and the failure of the state system to retain its monopoly on international violence. Emerging challenges come from both state and non-state actors adopting and employing unconventional methods to counter our advantages in traditional arenas.

Anti-Access Strategies and Capabilities

America's military capabilities are not measured in terms of the various capabilities owned or operated by the joint force, but rather by what capabilities the joint force can effectively bring to bear to accomplish its military or political objectives. Past U.S. military engagements in Iraq and the Balkans have relied on relatively close and safe regional bases from which combat operations could be launched. Potential adversaries view these regional bases as a critical necessity for the U.S. way of war. Future adversaries will increasingly attempt to limit, meter, or disrupt access to the local area of conflict. By developing the ability to limit and interrupt access to them, adversaries plan to degrade our military capability to a "manageable" level even if only for a limited period of time.

Future adversary forces will feature a set of integrated capabilities that are keyed to disrupt the ability of U.S. forces to close with the adversary and project power into a region. Furthermore, they will attempt to limit the ability of the joint force build, maintain, or communicate with regional power projection bases and complexes. They will target U.S. space capabilities such as optical, radar, and signals intelligence, and GPS navigation capabilities with laser systems and will use special-forces or long range strikes against ground stations. A second target for adversary anti-access forces will be the global information systems that U.S. forces require to synchronize and coordinate its forces. A third target will be regional and intermediate staging areas through saturation ballistic and cruise missile strikes, as well as other precision weaponry. Finally adversaries will seek to limit access to transportation nodes between the United States and the theater of conflict through attacks on railways, crane and port facilities, and mining of harbors.

Anti-access strategies will feature integrated political-military planning that integrates diplomacy, demonstration, deterrence, and coercion to keep other regional players from allowing basing and access to U.S. forces. Adversaries will likewise employ political and economic elements to limit or disrupt U.S. strategic deployment options. They will apply these tools to pressure possible U.S. allies and coalition partners to deny basing and overflight rights, eliminating the need for open military strikes. They will employ conventional munitions, weapons of mass effects, information operations, and combinations. As the perception of the inevitability of U.S. operations grows, exclusion will entail preemptive attack, possibly with weapons of mass effect. It is very unlikely that future adversaries will wait for U.S forces to position themselves regionally and await a set-piece U.S. onslaught as the Iraqis and Serbs have done in the past.

Opportunities
- Technical domains of air, sea, space, and cyber (essential to anti-access strategies) plays into traditional U.S. military strengths.

Emergence of New Terrorist Ideologies

Today, the globalized terrorist threat rests primarily on the emergence of a transnational violent and religiously-based ideology that has been captured and channeled by Al Qaeda. Although terrorists exist around the world to further various territorial and ethnic demands, only Islamic fundamentalism has managed to build a global non-state network of jihadist fighters, financiers, operatives, planners, propagandists and other ideologues to fight the United States and conduct operations against our interests. Other large-scale national, religious, environmental, and cultural ideologies, many with a nihilistic bent, may emerge and cross the line from political or social opposition and into armed conflict.

Citizens of the west may also take advantage of the terrorist warfare model and leverage the far superior economic and technical resources in the developed world. Operating in a relatively free legal environment that places a premium on personal liberty and privacy, these groups will wage disruptive warfare around the world on a scale that far exceeds that of relatively modest current day terrorist groups. How long would it take for a group of highly trained westerners to assemble nuclear or other weapons of mass destruction which include cyber or bio agents, and employ 'bleeding edge' technologies in unforeseen and devious ways? These 'Super-empowered' groups or individuals may create an even more elusive and dangerous adversary that seeks to cause disruptive conflict by attacking important cultural, economic, or physical infrastructures around the world.

Opportunities
- Widespread terrorism may limit state conflict and cause states to work together in order to reinforce and orderly and just state-based international environment
- Loose nature of the adversaries command renders them vulnerable to psychological operations and disruption of unity of effort.

Fourth-Generation Warfare Model

Fourth-Generation warfare is a model that attempts to explain a method of war that is developing around the world that responds to the western way of war and more specifically to U.S. doctrines of mass and maneuver land warfare and precision air attacks against critical nodes and infrastructure. [47] Fourth generation warfare reflects the notion that warfare is continuing to evolve, but is not evolving in terms of technologically-driven, U.S. led "transformation." Rather, and more ominously, fourth generation warfare is being built and explored by adversaries and its techniques have resulted in numerous victories. *[Note: The authors of this document do not necessarily concur that this model of generations of warfare is wholly valid. However, the construct does assist in understanding how emerging insurgency warfare differs from warfare that has been seen in the past as an aid to better forecasting implications and requirements for the joint force.]*

The fourth generation warfare model starts from the premise that state and non-state actors understand that they cannot directly face America's (and the West's) overwhelming conventional military power. Instead, these actors will rely on irregular methods of warfare that are explicitly designed to degrade and destroy America's political will. Fourth generation warfare idea is an evolved form of insurgency that makes use of all available networks – for example, political, economic, social and military networks – to convince the enemy's political decision makers that their strategic goals are either unachievable or too

---

[47] First generation is generally regarded as sociologically-driven Napoleonic War and levee en mass. Second generation warfare is a technological revolution applying industrial means to warfare (as represented by the First World War experience) Third generation warfare is seen as maneuver warfare that carefully arranges and exploits these technological and industrial advances through organizational and tactical means, and is represented by blitzkrieg and Airland Battle concepts.

costly for the perceived benefit. Fourth generation warfare is "rooted in the concept that superior political will can defeat greater military and economic power, and that these wars are lengthy often lasting decades."[48]

The fourth generation warfare model is not simply a tool of insurgency. Many nation states without the technological and sociological capacities of western powers will adopt fourth generation warfare methods to neutralize U.S. military capabilities. Irregular, un-modernized, adaptive forces with some conventional military capabilities as well as access to niche technologies will be the norm. In this way, all future adversaries will marry relevant high- and low technology based on the level that a particular society can support them, and will use them in ways that may not occur to the western mind. Some examples of this approach may include:

- Improvised explosive devices with explosively formed penetrators, commercially-derived (and cheap) sensors, and biological/radiological packages.
- Advanced small arms enhancing sniper operations and providing an anti-materiel, armor piercing capability.
- Enhanced indirect fire munitions increasing effectiveness through use of proximity fuses, cluster munitions, and smart munitions.
- Improved Anti-tank guided missiles and rocket-propelled grenades used against a variety of hardened point targets, and not exclusively armored vehicles.
- Proliferation of low altitude air defense and surface to air missiles married to sophisticated context-specific ambushes.
- Thermobaric weapons.
- Increased use of unmanned aerial vehicles in both ISR and attack roles.

Ultimately, a "fourth generation" opponent understands that victory on U.S. terms is not possible, but that their chances of winning are improved if they can wear down the political and societal will of the U.S. to engage. The form of warfare relies on ready access to powerful technologies, a will to use relatively unrestricted violence when required, and the development of cultural and identity-based networks to carry on the fight. Because a fourth generation adversary has very little bureaucracy or requirement to tell the truth, it is able to adapt and shape perceptions about conditions on the ground very quickly. This agility, coupled with access to the global media and the ubiquitous availability of cheap and powerful media tools, allows them to freely maneuver within the media environment and magnify U.S. mistakes or inject false propaganda almost at will.

Opportunities
- Understanding "4th generation warfare" would foreclose opportunities of adversaries to compete with or oppose the United States.

Disruption of Global Trade and Finance

The global trade and financial network is a key source of power for the United States, and defense of the global trade and finance regime, as well as key nodes that underpin the

---

[48] T.X. Hammes. *The Sling and the Stone* (2006).

international trading networks may be a central element of U.S. national security strategy. Potential adversaries understand that the U.S. economy is a potential center of gravity that may be vulnerable to attack and disruption. The internet reaches deep into all parts of American society, allowing adversaries to influence anything connected to it. Millions of shipping containers enter a few major hubs, either providing the means to move material into the U.S. homeland, or providing a key node that might be damaged by a determined adversary. Any disruption of the flow of oil products, refining capacity or port traffic would have a significant negative impact on our economy, and many private entities, in the trade and finance area – such as the New York Stock Exchange – may be vulnerable to cyber attack. Vulnerabilities in the trade and finance regime and the collapse or retrenchment of the system – in addition to decreasing the economic well-being for its citizens – may spill into the military domain and be the source of surprise as the joint force commander may be called to protect global trade or finance nodes and networks.

These interlocking financial channels in theory, lead to a more efficient allocation of capital to investments worldwide, But the increasingly complex financial mechanisms also transmit trouble: For example, in 2007 what began as a problem in one sector of the U.S. housing market – mortgages for borrowers with poor credit – has infected credit markets worldwide. The latter problem was created when global financial institutions repackaged these mortgages as sophisticated securities and sold them to banks, corporations, and local governments around the globe. As a result, local price and liquidity shocks are very likely to spread around the world and create havoc for capital markets around the world. Additionally, any economic weakness in the U.S. can hit other countries by unsettling global financial markets and curbing access to capital and depressing trade.

Opportunities
- The U.S., with its large and diverse internal market may be more able to weather trade or financial disruptions than other countries around the world.
- The U.S. economy is highly flexible, and weathered significant setbacks, such as the collapse of large hedge funds, and the attacks of 9-11 with minimal damage to the larger economy.

Persistent Cyber-Conflict/Disruption of Information Networks

The continued and rapidly increasing expansion of information technology and systems will greatly assist commanders and other actors. Complicated networks of landlines, radio relay stations, fiber optics, cellular service, and the Internet provide massive communications capabilities that are used by governments, businesses and individuals around the world. As more critical activities, including the remote control of infrastructure systems and the movement of money and finance ride on these networks, they have increasingly become the targets of manipulation and even disruption from criminal enterprises and both state and non-state adversaries. Eventually, these information networks may become so insecure and vulnerable that they become difficult or impossible to use.

A number of adversaries are beginning to understand the use of the internet and information warfare in this domain. In 2007 an unidentified group conducted an information blockade against Estonia in which government and financial communication systems were cut off from the outside world. This attack isolated the country as effectively as a naval or land blockade of the country and inflicted substantial financial damage, and demonstrated the inability of governments to address this type of attack. In the future, well organized criminal groups, state, and non-state actors will command millions of individual computers across the internet and harness them in difficult to trace "botnet" attacks against adversary systems. Even today, a criminal organization commands the "Storm" botnet which represents the greatest accumulation of computer processing power on earth. To what ends will this botnet be used, and how will the military address conflicts with organizations armed with these types of capabilities in the future?[49]

Opportunities

- U.S. is a leader in internet technology and owns the domain-name registration system. This advantage may be used in the future to severely disrupt adversaries that use internet capabilities.

Proliferation of Weapons of Mass Destruction or Effect

The proliferation of the ability to kill or injure large numbers of people for a small or inconsequential level of technological or financial investment will be a central challenge for the future joint force commander. Weapons of mass destruction will become an ever more challenging and multifaceted problem as "warhead" become more compact (down to the size of self-replicating viruses), and may be delivered from a wide variety of platforms from the standard ballistic/cruise missile combination, to infectious agents or pathogens carried by an international traveler. Globalization, access to information and global travel, and economic factors will are producing conditions where the use of WMD/E against the US and our allies is not only possible, but increasingly likely. A number of states will begin to develop nuclear deterrents of their own and employ them to dissuade the United States from becoming engaged in their regions. A number of more specific challenges within the area of WMD/E proliferation include:

- **Dual-use technologies.** Future WMD/E technologies will have many similarities and synergies with legitimate and beneficial scientific, technical, and economic endeavors. Nuclear energy will be a more prominent feature as fossil fuels become rarer. Bio-engineering of vaccines, chemical fertilizers and pest control are all key parts of human social and economic well-being. Adversaries will have the ability to pursue small WMD/E research and development programs under cover of legitimate programs. Biological laboratories have especially small (and decreasing) footprints and their dual use nature makes detection and elimination of a weaponized program particularly challenging.

- **Uncontrolled WMD materials.** Potential weapons grade nuclear material is abundant, poorly controlled, and insufficiently secured. This material ranges from actual warheads in the former Soviet Union to spent nuclear fuel stored in multiple locations around the world. Some of this material may be at risk from criminals seeking to steal and sell it to potentially hostile states or (more likely) international terrorist groups. Though there are protocols attempting to control nuclear substances and technologies, no such conventions exist for much of the material

---

[49] Sharon Gaudin, "Storm Worm Botnet More Powerful than Top Supercomputers" *Information Week*, September 6, 2007.

necessary to develop and produce biological and/or chemical weapons and weapons of mass destruction/effect.

- **Concealment/detection.** An adversary's ability to conceal WMD/E is outrunning our ability to detect them. Perhaps the greatest challenge to the detection of WMD/E is the sheer volume of space available to an adversary compared to the actual size of a WMD/E device. Small nuclear weapons are a current reality. The size of chemical or bio agents needed to produce a mass effect can be negligible, and both may be hidden among the vast community of international trade and travel. While technological efforts to detect these devices and substances and implementation of policies to deny their entry are underway, significant shortfalls are likely to persist over the next 25 years.

- **Constraints on use.** For nation-states, the anticipated retaliation and universal condemnation have served as a deterrent to indiscriminate first-use of WMD/E. Two emerging phenomena are loosening those constraints. One is the emergence of nihilistic non-state actors, such as Al Qaeda, who offer small, low-value targets for physical retaliation and have no reluctance to employ WMD/E against the United States and its Allies. A second phenomenon is the lack of a forensic signature for many types of WMD, enabling their use without a reliable, verifiable way to attribute their origin – the anthrax attacks against the U.S. immediately following 9-11 is an example of this. Combine these two and it is very possible to envision an adversary conducting a WMD/E attack with relative impunity and, to him, acceptable risk.

Opportunities

- Genocidal nature of these capabilities may encourage the development of wider international consensus and active cooperative efforts to limit the proliferation of WMD/E.

Failing Nuclear and Energy States

A special class of failed state is that of states armed with nuclear weapons or those with significant oil production and export capacities. Both a failed nuclear energy state and a failed energy state would have significant implications for U.S. security. Future forces will be faced with the prospect of missions to either prevent nuclear weapons or materials from being lost and controlled by non-state terrorist groups, or falling into the hands of hostile factions within the state. The collapse of an energy exporting state may draw in U.S. forces that may be tasked with controlling and operating production facilities and ensuring that supplies of critical hydrocarbon resources are available to the world economy.

Opportunities

- Success in maintaining access to the global energy supply on behalf of a tightly interdependent economic system will burnish America's leadership role.

Failed Mega-City

As the world's urban population grows to 4.6 billion over the next 20 years, the stresses and strains on national and city governments will become intense. Some of these cities will fail, and massive human disasters may result. It will become increasingly difficult to avoid military operations in an urban environment, and U.S. forces may be drawn into these areas to support human populations as well as conduct operations. Future adversaries will use these failed urban areas to negate standoff, mass fires, and sensor capabilities and create strongholds where opponents can achieve sanctuary U.S. military activities. These forces will capitalize on the nature of cities and their populace, as well as open source data for the information needed for decision making. They will seek to cause heavy casualties and collateral damage to influence the will of the U.S. and its coalition partners, while trying to win the hearts and minds of the given society.

Urban environments typically feature subterranean infrastructure, shantytowns, and skyscraper canyons in varying states of functionality and repair. This complexity can degrade or reduce mobility, as well as the effectiveness of high-technology weapons, communication systems, and intelligence, surveillance, and reconnaissance (ISR) capabilities. In addition, population density effects countless complex social and cultural interactions that influence human intelligence and open-source information while increasing the risk of collateral damage; e.g., more civilians are likely to be harmed or killed.

Figure 3.2. Urban Terrain in the Third World

The U.S. will be faced with difficult challenges of conducting operations within these environments – attempting to separate the local population from supporting or being held hostage to adversary forces. This will require a number of sophisticated capabilities including human and cultural mapping, running of city services and utilities, while conducting very precise combat operations against opposing forces. Furthermore, it will occur within urban agglomerations that sprawl across hundreds of square miles and contain tens of millions of people. Frequently, small combat operations teams that combine warfighting, police, and civil affairs capabilities will be present in the environment as adversaries, allies, or neutrals. The opportunities for close contact in this environment will multiply force protection requirements.

Opportunities
- Ability to stabilize a failed mega-city could encourage positive views of the United States.
- Training for failed-mega-city stabilization could be useful in humanitarian disasters or WMD/E remediation activities.

Global Anti-American Coalition

A likely future U.S. strategic goal will be the expansion and reinforcement of global relationships in order to ensure that key states in the international system have a greater stake in participating in that system than in fighting it. Such as strategy is based on ensuring that key states around the world are closer to the United States (or have more in common with it) than any other major power in the system and to avoid a balancing coalition against it. In order to maintain a "U.S.-centric" international system, it must hold together and balance the numerous interests of its allies, friends and partners, while increasing tying possible competitors into that system. Bismarck's Germany of the late 1800's tried a similar feat in Central Europe (although on a significantly smaller scale) and could manage these complex and difficult arrangements for a time – however, his system collapsed into grinding and annihilating war not long after his death. All leading nations are challenged at one time or another as their power declines relative to others. The balancing and shifting required to sustain the current U.S.-centric system may be too complex for our political system and our national instruments of power to maintain.

Over the next 20 to 30 years, the system of international relations developed by the United States after the Second World War and expanded after the collapse of the Soviet Union may be challenged by the emergence of a global anti-American coalition. Such a coalition would be comprised of states with significant disagreements or conflicts with the United States but would also include a disparate partnership of anti-American groups and organizations dedicated to minimizing U.S. power and influence around the world. For example, in parts of the Arab Middle East today there is a strong measure of anti-American sentiment because of the ongoing Palestinian conflict, our presence in Iraq, and the perception of "anti-Islamic" misdeeds in Guantanamo Bay. Already in 2007 the Shanghai Cooperation Organization (SCO) and other treaties between Russia, China, and the Central Asian states have been signed and are designed to limit U.S. activities and access in the region. Other states, such as Iran and Venezuela may be added to such SCO-like groupings and could become more actively opposed to America's influence and position around the world. The aftereffects of Soviet Union's Cold-War disinformation campaign against the United States may metastasize and be amplified by anti-American groups and radical Islamists around the world, including a ready core of transnational leaderless networks who will work to degrade and attack the United States in both the information and physical domains.[50]

Opportunities
- A visible and active anti-American coalition may encourage some states to develop closer security relationships with the United States.
- Freed from major Eurasian positions, the U.S. may rely on a more mobile, global security posture based on its dominance of the seas, air, and space.

3.3 National Security Shocks

This topic refers to the idea that we could undergo a significant period of discontinuity in both national and international affairs. Scientific progress is accelerating so swiftly planning for technological change for 2030 today is like planning for 2007 in 1880. Complex systems such as our energy infrastructure, global and national economy, and technological supremacy are under stress as new and powerful actors take advantage of the global economy. Our inability to defend against the September 11th attacks on New York, Washington, and

---

[50] See Andrew, Christopher, Vasili Mitrokhin (2005). *The Mitrokin Archive II: The KGB and the World.*

Pennsylvania were less a failure of capability than a failure of imagination.[51] This section of the JOE will apply imagination to explore a number of unlikely but highly consequential challenges to our nation's security and the role of U.S. military forces to address them.

Energy Disruption

The security of energy supplies which are essential to the U.S. economy is a potential source of dramatic change – either positive or negative – for America's security posture. Currently, the U.S. economy is based on a highly inelastic demand for one resource – oil. The supply of oil imports may be threatened by determined adversaries at a few vulnerable geographic locations or facilities. For the next twenty years the world's energy supply will remain largely based on petroleum, which will be constrained and will continue to originate from highly volatile locations – the Middle East, sub-Saharan Africa and Central and South America. Al Qaeda's leadership stated its desire to disrupt the West's access to oil and to destroy the Saudi royal family, and Iran practices for this mission on a regular basis to enable it to either destroy critical Saudi oil facilities or interrupt the flow of oil through the Straits of Hormuz.

Given the fragility of our crude oil supplies over the next twenty years, expanding world consumption, and the lack of meaningful large scale alternatives to oil, access to and source protection of oil sources will continue to be a major policy and security focus of our nation and our competitors well into the future. Thus, the joint force commander must consider the intersection of these trends. The key questions for America are: How vulnerable is the supply of oil upon which our economy is based and; do viable alternatives to oil exist? The U.S. has benefited from a world oil market that is generally open to investment and trade, and at least partially based on market forces. The emergence of an international energy regime where suppliers and users lock in guaranteed supplies (and conversely, "lock out" access for the United States) could be a cause of future international conflict.

Opportunities
- Increased price of fossil fuels encourages the development of alternative energy sources, which minimize American dependence on volatile foreign sources of fuel.
- Energy production increases domestically, lowering the need for imported energy and decreases revenues for hostile and aggressive energy states.

Technological Surprise

The U.S. share of world investment in research and technology are decreasing as new powers rise and the U.S. economy shrinks relative to that of the rest of the world. Access to technology on the global market is slowly eroding the historic technological primacy that the US has enjoyed, and others will increasingly develop niche military capabilities that may be unanticipated by the scientific and defense communities. The future joint force commander may fight in an environment where technological capabilities of an adversary are less well understood and may work to negate some key U.S. military capabilities – much as the U.S. development of stealth negated massive Soviet investments in air defenses.

Opportunities

---

[51] *The 9/11 Commission Report*, Chapter 11.

- The development of a strategic S&T scouting effort linked to the U.S. university and private R&D communities may allow the U.S. to exploit "leapfrog" technologies developed elsewhere.
- Challenge of new technological powers may encourage greater investment on science, technology and engineering education in the United States.

Nuclear Attack

With the widespread availability of sixty-year old nuclear technology becoming more widespread, the challenge of preventing an attack and/or the recovery of one or more major U.S. cities after such an attack will remain a challenging problem for the joint force. The detonation of a nuclear device against the United States or its allies will have catastrophic implication for the international security regime. A nuclear attack may take place inside the United States through a smuggled device across relatively unguarded international borders, or may be inserted into a shipping container. It may also take the form of ballistic and cruise missile attacks, either at intercontinental distances, or based on submarines or ships sailing off either coast.

Opportunities
- The U.S. superiority in nuclear weapons and delivery means allied with our technological ability to trace the origin of nuclear material will deter most nation states from the use of their nuclear weapons or the clandestine delivery to other radical actors.

Pandemic

A pandemic disease is one that is readily transmissible through human populations, kills or injures those who are infected, and rapidly spreads regionally or globally. Pandemic diseases may be naturally occurring, as with the case of the 1918 flu epidemic or may be intentionally spread through, for example, weaponized smallpox virus. A pandemic in North America would be protracted and pervasive, causing substantial societal impact and persistent economic losses in almost every state. A typical annual flu results in 200,000 hospitalizations and 36,000 deaths. A large-scale pandemic influenza event in the U.S. could result in 75 million hospitalizations and as many as 1.5 million deaths. A further extrapolation of the 1918 event to the current day would result in the lost of 77 ship crews, 10 Army brigades, and the loss of one of three Marines divisions.

A biological attack to induce disease presents the ruthless adversary a ready weapon of mass terror to disrupt the American economy and society. A smallpox or artificially-build biological weapons has the capacity to kill many more people than a nuclear attack. Further, the proliferation of biotechnological capabilities is makes diseases like smallpox will relatively inexpensive to produce and difficult to detect until released. These techniques also mean that terrorists groups or other adversaries may have the capability to modify existing disease to be more lethal and/or more transmissible. If the smallpox is injected directly into "suicide" volunteers, these volunteers become both the storage and dissemination systems. Using a few volunteers and commercial airlines, a terrorist group could create a near simultaneous worldwide outbreak of smallpox.

Joint forces will certainly have a role to play in a dramatic pandemic event and may be required to provide certain services assist both the Nation and the world in the prevention and mitigation of such a naturally occurring event or deliberate attack.

Opportunities

- The threat posed by pandemic diseases provides opportunities to train and exercise with other nation's militaries and governmental and non-governmental civilian organizations to establish common processes, procedures, and methodologies through a larger system of collective security and engagement.

Global Depression

A global depression or other severe economic retrenchment resulting in the loss of U.S. economic preponderance would have dramatic effects on America's national security posture. A substantial reduction in U.S. or world economic activity might be triggered in a number of ways. For example, in the U.S. our strong economic base could be hurt by a rapid decline in productivity growth, a significant loss of market share of U.S. capital markets, or a material decline in education standards, especially in math or science. The U.S. would be the source of the most dramatic and painful consequences as a result of its position as the world's largest economy and an important world destination for exports. Expansive trade and current account deficits, significant amounts of national and private debt held by foreign individuals and countries, and a resulting collapse in the value of the dollar would severely impact the ability of the U.S. to meet its global military and financial commitments. China may also be the source of economic disruption should its capital markets collapse because of over-speculation and the inability of the Chinese Central Bank to assure steady macroeconomic growth. Additionally, China may also face economic disruption if there is a halt or retrenchment in the tremendous rate of growth in Chinese consumerism. Failure of economic globalization may result in increased regionalism, and greater tensions between economic centers as globalization gives way to increased trade restrictions, mutually-destructive currency competition, and the potential collapse of the Euro.

Opportunities

- This may be considered a "worst case scenario" for U.S. national security, however, as bad as global economic collapse, depression, or a retreat from open trading standards might be, the U.S. is better-positioned to weather this condition than many other states. It has a large and unified home market to fall back on, as well as a relatively well-educated population with significant financial resources and physical resources to fall back on. The U.S. may be weaker, but others may fall by even greater margins.
- Most of the world has an interest in the value of the dollar and would incur economic damage by its collapse. This makes an attack on the dollar a risky strategy to anyone capable of carrying out such an attack as it would bring about potential shocks through the global monetary system as the value of the dollar is degraded.

Loss of Access to Portions of the Global Commons

The global commons are environments that are outside of national jurisdiction, and provide access to much of globe. This definition includes natural areas such as the oceans and airspaces above them, the Antarctic, earth orbit, outer space and celestial bodies such as the moon, but may also include "space" within the connected global communications networks such as the internet. [52] Airspace technically belongs to the countries that lay beneath it, but few can exercise de facto control of it above 15,000 feet and thus may be similarly included.[53] Access to and subsequent dominance of these commons is central to the global U.S. position, they are central to our ability to access the world and our control of them allows the U.S. to cut others off from the world in times of crisis and war. Command of global commons such as the high seas allows the United States much room for maneuver and can significantly deny similar room to adversaries.

> *"...the U.S. has relied on satellites, air superiority, and immunity for its rear area facilities and operating areas, including the sea. It was challenged in the maritime commons by mines and the threat of air and cruise missile attacks in Desert Storm, but has not been threatened since."* [54]

Command of the commons is enabled by "all the difficult and expensive things that the United States does to create the conditions that permit it to even consider one, two, or four campaigns."[55] For example, the U.S. has massive investments in satellite reconnaissance, communications, and navigation which allow it to understand the world to an amazing degree. However, this reliance on space means that potential opponents will seek to deny our access to space systems that allow the joint force to tie together globally-ranging military capabilities. Likewise, the U.S. may lose the ability to command the sea, air, space and internet. Rather than reaching out to the world, our adversaries could have the ability to reach in to the United States and to isolate us from friends and allies around the world.

Opportunities
- Challenges to U.S. control of the commons may encourage the U.S. to redouble efforts in this area. For example, directed energy systems may allow the U.S. to expand its command of the air to deny any activity in the air – including ballistic missiles, artillery, mortars, and other airborne systems.

---

[52] Richard Fernandez, "The Last of the Global Commons" March 29, 2007.
[53] Barry Posen, Command of the Commons: The Military Foundation of U.S. Hegemony, *International Security* Volume 28, No. 1 (Summer 2003). p. 8.
[54] Frank Gaffney, *The American Way of War through 2020*. p. 12.
[55] Ibid, p. 7.

Chapter 4: Implications for the Joint Force

Chapter Four of the JOE document develops a set of more specific joint force implication based on the trends and potential military challenges described in Chapters Two and Three. These implications, while neither inclusive nor in depth, serve as a point of departure for further discussion throughout the futures enterprise and are meant to assist in the shaping of concept development and experimentation efforts throughout the Department of Defense. The periodic and iterative nature of the JOE dictates that these implications will be further refined as environmental trends and military challenges evolve.

These implications are placed into five general categories reflecting key considerations of joint force commanders at the operational level of conflict. These categories are terrain, base, knowledge, force application and command and will be discussed in further detail below. Together these five categories encompass enduring factors that a joint force commander considers when planning for or conducting battles, strikes, and campaigns in support of national strategy.

4.1 Terrain

The concept of terrain includes the physical geographic space upon which conflict occurs. The notion of terrain also includes the mental, moral, cultural, and societal dimensions that frame the arena of conflict. Terrain defines the physical and intellectual context of conflict or war. The future operating environment has several specific military implications based on the idea of changing or expanding notions of terrain. These are:

- The Unified Global Battlespace
- Position and Influence in Human Terrain
- Comprehensive Approaches

The Unified Global Battlespace

The size of the battlefield continues to expand as the ability to find, track and kill targets grows. As a consequence, the historic trend of decreasing density of combatants per meter of "front" continues as the tools to communicate, influence, and move allow the individuals ever greater reach. Figure 4.1 below illustrates this dramatic historic trend. [56] Today the battlespace is approaching global dimensions, while the effective range of influence available to individuals can span thousands of miles. For example, regional terrorists may transmit video footage of a convoy attack to a news organization that then beams the footage across the globe. On the other hand, an unmanned aerial vehicle (UAV) operator can direct a response team to capture or kill the very same terrorists from a base thousands of miles away in Nevada.

As forces continue to disperse, and their ability to influence, disrupt, or destroy targets expands, lines and "fronts" of battle become complex, three-dimensional "volumes" and units operate in small, dispersed units linked by ubiquitous communications capabilities. [57] Operations by these units will occur around the world at multiple locations simultaneously

---

[56] Chart from Russell W. Glenn, *Heavy Matter: Urban Operations' Density of Challenges.* RAND Corporation Monograph, 2000.

[57] Laurent Murawiec, Innovation, Element of Power. (Geopol CASE), p. 26.

while their tactical activities will be linked to other tactical activities at multiple locations across the globe in the air, on the ground, at sea and in space. The United States must prepare its leaders, command and control apparatus, and organizations to withstand the simultaneous assault of multiple domains at multiple locations around the world. The national security community must come to grips with wide ranging activities by adversaries that are highly capable and adaptive.

Battlefield Density through the Ages

	Antiquity	Napoleonic	U.S. Civil War	WWII	1970s
Urban Examples	Plataea New Carthage	Jungingen Aspern-Essling	Monterrey Churubusco Rorke's Drift	Stalingrad Aachen Manila Berlin	Beiruit Khorramshahr Hue
Combatants Per km²	100,000	4,970	3,883	32	25
Urban Examples	16,300	46,400	11,600	1,300	1,100
Km² Per Combatant	10	201	258	31,000	40,000
Urban Examples	61	22	86	769	909

Figure 5.1. Battlefield Density

Adversaries will attack friendly forces and the U.S. military will be taking the fight to them around the world and in all domains – air, ground, sea, space, and information. From a wider perspective, the military has to broaden its perspective on modern conflict to simultaneously manage many fights—some within the United States itself, some outside, some tangible, some technical, others still intangible and cognitive in nature.

Position and Influence in Human Terrain

The future environment will require joint forces that are capable of mapping and surveying human terrain, and then developing capabilities that allow our forces to identify positions of advantage and to position our forces to influence them. Examples of an advantageous position in human terrain may be a tribal or social leader outside formal government structures who can influence a large swathe of the population. Influence may include a wide range of inspiration, leadership, pressure, suggestion, persuasion, or force that causes them to behave in a manner consistent with the commander's intent. When an adversary is fighting on his home ground, he will take every opportunity to aggravate and intensify the friction that occurs when two differing cultures must interact; in the context of an emerging or ongoing conflict, he can leverage this friction to his advantage. Adversaries will exploit every opportunity to publicize any real or perceived assault on the native culture by the U.S. Their success will enable them to intensify opposition to the U.S. and extend the duration of the conflict.

The key implication is that the joint force to acquire cultural expertise and the ability to gather culturally relevant information. This cultural knowledge must be integrated into operational concepts and plans to allow U.S. forces to more effectively gain the cultural initiative. The U.S. has demonstrated the capacity to deal with the military and territorial challenges of a conflict. We have also wielded diplomatic and economic power to influence an adversary's political and economic systems. A military implication of the future joint operating environment is to understand and relearn the skill of winning the cultural component of a campaign. A successful cultural campaign will work directly to defeat the adversary's will to continue the conflict which has the follow-on effect of being a central component post-conflict stability.

Comprehensive Approaches

The future joint force will operate within an extended battlespace that includes political, military, economic, social, and other elements which will underpin an adversary's will and ability to fight. Furthermore, a skillful adversary will operate in each of these domains against the United States. How the joint force, adversary states or a transnational terrorist force influences, denies, disrupts, or destroys these elements or the essential linkages between them will defines the winners losers in future conflict. The joint force must understand how to operate with other elements of government and society to affect the adversary. It must understand critical vulnerabilities in other domains over which the joint force might not have exclusive control. Finally, it must work with other organizations in the planning phase of day-to-day operations in order to influence world events to avoid combat or to influence the course of conflict should it occur.

4.2 Base

A base from a physical standpoint is a locality from which operations are projected or supported, or an area or locality containing installations which provide logistic or other support.[58] From the mental or moral standpoint it reflects an ability to gain strength through sources of legitimacy or a common ideology and provides the foundation from which force or influence can be directed against an adversary or enemy. A base is both a position of strength and a point of balance from which force is applied. The notion of a base is analogous to Clauswitz's concept of center of gravity, which was described by one military analyst as: "the *focal points* that serve to hold a combatant's entire system or structure together and that draw power from a variety of sources..."[59] Ultimately, a base is a physical or intellectual place in which the force can gain strength or is the source of legitimacy for its actions.

---

[58] Joint Publication 1-02. *Department of Defense Dictionary of Military and Related Terms.* 12 April 2001, as amended through 17 October 2007.
[59] Dr. Antulio Echevarria II, "Clausewitz's Center of Gravity: Changing our Warfighting Doctrine – Again!" (September 2002)

- New Sources of Strength
- Secure Access to the Global Commons
- No Sanctuary

New Sources of Strength

The United States will likely remain the dominant economic, military, and political power during the next twenty-five years. Other actors will accumulate power in certain dimensions and some, perhaps, will directly challenge the United States. It is unlikely, however, that a single nation or coalition will be able to match the power of the United States in its entirety (except in the case that the national security shock *Global Anti-American Coalition* outlined in Chapter 3 above emerges). Yet, a clear diffusion of power is underway within international relations. For example, the spread and application of technology—specifically information technology—is resulting in the diffusion of power away from central governments. More power is being distributed on supra-national and sub-national levels as well as outside of the state system altogether than ever before. Centralized state control is no longer the norm and these shifts will affect how the United States relates to other actors.

Powerful transnational threats and the impact of autonomous regions erode state control and scope of direct diplomacy, while global criminal networks with access to vast wealth threaten stability of states, economies, trade routes, and other key features of the international environment. Because opponents are diverse and often below state level, the joint force commander must understand the foundations of their ability to project power and influence, and have at his disposal the capabilities to destroy, degrade, or otherwise deny their ability to leverage new sources of strength.

Secure Access to the Global Commons

The global commons, principally the high seas, outer space and cyberspace, are domains in which mastery of technical knowledge and operational procedure has a disproportionate effect on ones ability to control it. Because the U.S. will remain a prominent technical power and has a wide margin of advantage today, a key implication for the joint forces is to ensure the protection of our dominance of the high seas, space, global networks, and the airspace above fifteen-thousand feet. The global commons can become a key protected feature from which U.S. forces might be based, allowing the joint force commander to position and maneuver to key points of advantage and to conduct influence or fires from the global commons in times of conflict and war.

No Sanctuary

Very few U.S. strengths or bases will be sanctuaries shielded from the threat of attack or disruption by a determined adversary. Adversary forces will conduct attacks within the United States, focusing on the disruption of strategic deployment assets and methods, including military installations, lines of communication, and sea and aerial ports of embarkation. The adversary will also conduct attacks at national command and control facilities. All will be conducted multiple locations and use a variety of weapons including weapons of mass effect, non-lethal capabilities such as electromagnetic pulse weapons, and will attempt to reach into and disrupt U.S. networks using electronic and information warfare methods. The loss of sanctuaries means that planning must change to include military families, key civilian knowledge workers, and owners of critical infrastructure that support department of defense bases and lines of communications. This planning may also include local law enforcement officials and organizations, and state and local officials who operate emergency operations centers.

4.3 Knowledge

Knowledge describes the ability to gather and use information for purposeful action in a conflict. The joint force must have the ability to engage in the acquisition, production, and dissemination of knowledge to understand the nature of conflict and adversary, and to direct the application of force.

- Information Systems
- Knowledge Parity
- The Knowledge-Based Organization

Information Systems

Knowledge is critical for making decisions faster and better than the adversary and for sustaining the advantage of knowledge and decision dominance. But, because it faces smart, adaptive, learning adversaries, the U.S. military must understand that the conditions of superiority and dominance will be severely and continuously contested. Adversaries will wage a "knowledge war" over valuable knowledge – physically and in cyberspace. The principal tool to wage this war will be information operation. The future joint force will encounter adversaries that place the majority of effort on the information campaign, while supporting the information campaign with violence only where required – but with a high degree ferocity and utter lack of restraint.

Information systems aid us in our ability to fight for knowledge, however, the more we rely on information systems, the more likely adversaries will be to attempt to misdirect, disrupt, or destroy them. The emergence of electromagnetic pulse weapons may create severe difficulties for a "network centric" view of warfare. Thus, the joint force must understand how to most effectively use information systems, must balance the uses of networked systems against the need to harden and defend them, and must prepare to operate in environments where they are denied or otherwise rendered ineffective. The future joint force should incorporate knowledge war into its thinking, lexicon, doctrine, and training, and modeling and simulation capabilities should ensure that the information domain, knowledge environments, and communications architectures are incorporated in training and mission rehearsals.

Knowledge Parity

Adversaries will have much the same access to useful and valuable data, information, and knowledge as the United States. The information-rich environment described in this document significantly levels the playing field for adversaries who can take advantage of billions of dollars of commercial investment in imagery and communications technologies. In an environment of information equality, advantages will most often flow to side able to adjust their organization to take advantage of ubiquitous information and act on that information more quickly or precisely than others. What is more important for a cultural fight: the information gathered by a small group of operators in the field, or the trillion-dollar investment in remote sensing and surveillance?

Some of this information and knowledge will come from collection operations; some will come from open sources such as television news and open source databases. Perhaps the most difficult aspect of information to control will be commercial intelligence. The explosive growth of high-quality satellite and aerial imagery may be paralleled in the future by growth in commercial human intelligence, measurement and signature intelligence, and signals intelligence for commercial purposes. Adversaries will also use commercial knowledge or intelligence analysis including knowledge product research and packaging. In addition, there will be an array of commercial business intelligence tools and databases available from the internet. The processors and databases needed to engage in data mining and the very quick production of trends, relationships, and places of interest contained in available data, information, and knowledge will be for sale to anyone. The U.S. military must train to operate in an environment in which its adversaries have access to high-grade and timely data, information, and knowledge such as that listed above. Military trainers and educators have to replicate such adversaries with their intent, money, and access to valuable data, information, and knowledge.

The Knowledge-Based Organization

Knowledge will be created throughout the world, both inside and outside military systems. A knowledge-based organization must understand how to access the vast array of information available throughout the world, connect experts, planners, and operators, and translate the knowledge gained in these interactions into military operations. An implication for the ubiquity of knowledge is that the procedures and tactics for moving information around the battlespace must reflect the need to share hard-won human intelligence among operators, and upward to strategic and national planners. Likewise, intelligence support must be organized to supply national capabilities to tactical units, allowing the downward flow of important "national technical means" to effectors wherever they are. Furthermore, self-publishing capabilities and advanced search mean that high-quality analysis will occur in often unlikely places.

Future adversaries will attempt to find and attack critical links, nodes, seams, and vulnerabilities in U.S. systems that offer the best opportunity to "level the playing field." They will conduct some variant of effects-based operations, planning, and assessment against U.S. activities and systems. This entails ISR capabilities linked directly to fires (tactical, operational and strategic, lethal and non-lethal), tailored operational formations, paramilitary, special-purpose, and guerrilla units all tasked to affect specific capabilities whose loss or degradation will significantly reduce overall force effectiveness.

4.4 Force Application

Force application refers to the activities conducted by parties to a conflict with the goal of changing the behavior of others. In warfare, force application refers to the sum of maneuver and fires that the joint force commander brings to bear against an opponent. Fires are the use of any weapon systems to create a specific effect on a target, while maneuver is the employment of force across terrain (often in combination with fires) to achieve a position of advantage with respect to the enemy in order to accomplish a mission or to throw the adversary off balance. [60] In other conflicts, (economic conflict for example) the subsidiary elements of the application of force analogous to maneuver and fires includes the concepts of *position* and *influence*. Position is the arrangement of assets in an advantageous place, such as the purchase of significant foreign exchange reserves to resist trade sanctions. Influence is the use of assets (analogous to "weapons" in our military example) at ones disposal (including intangible cognitive skills such persuasion or discussion) to create effects against a target. The practical results of the application of force are that an adversary's base can be damaged or disrupted or his ability to use or traverse terrain is denied.

- Broad Definition of Military Capabilities
- Innovative Use of Emerging Technology
- Cognitive Campaigns
- Lawfare

Broad Definition of Military Capabilities

Effective strategies by state or non-state actors will attempt to synchronize activities in the diplomatic, economic and information domains, in addition to military actions create the greatest tangible effects on adversaries. For example, a traditional attrition-based perspective on warfare sees a "target set" of tanks, aircraft, artillery pieces and other discrete units of military power; a broader definition of capabilities understands dynamic flows of material, ideas, money, or other element of power through systems. Combat power as such, is not simply the specific platforms that make up a military force, but the also the logistics and economic systems that keep it supplied, the command and control that keeps it oriented, and the political and social will that keeps it directed. Viewed from this perspective, what may look from a simple attrition/maneuver perspective like a formidable military force (pre-1991 Iraq) may be a fragile tool of statecraft indeed. Likewise forces that may seem weak and barely above the "military horizon" may be able to "asymmetrically" conduct formidable operations with dramatic strategic effects – the attacks of September 11th 2001 being a prime example. [61] The implication for the joint force is that it must be capable of integrating military activities with social, political, diplomatic, economic, and other activities in concerted action.

The military dominance of the United States and its likely allies will force many potential adversaries to consider alternatives to conventional warfare. Ready access to information technologies will empower militarily insignificant adversaries to gain the knowledge to apply an effects-based approach toward any conflict with the United States. We can anticipate any adversary will attack our systems with whatever degree of sophistication he is capable of. As such, the future operational environment will encompass not just our military, but the United States homeland infrastructure and the political, economic, social, and information systems that form the basis for our national power. A full scale conflict with a great power competitor would require a high degree of knowledge to damage or disrupt adversary's systems and defend our own systems from a sophisticated attack.

---

[60] Fires and maneuver definitions derived from Joint Publication 1-02. *Department of Defense Dictionary of Military and Related Terms.* 12 April 2001, as amended through 17 October 2007.
[61] The "military horizon" is outlined by John Keegan in *A History of Warfare*

While the adversary will not attempt to hold ground or conduct combined arms warfare against U.S. territory, warfare will be conducted as strikes that focus on the erosion of national will by violent attacks against civilians, and to disrupt our ability to fight abroad by attacking critical nodes in those systems vital to support military operations. Over time, the adversary hopes to cause the withdrawal or disengagement of opposing U.S. forces and capabilities– without having to actually engage and defeat our military forces in the field. All capabilities will be aimed at exhausting the U.S. strategically over time.

They will avoid fighting U.S. forces in less complex and open environments that favor U.S. standoff technology, precision guided munitions, and intelligence, surveillance and reconnaissance capabilities. Adversaries will seek to use complex terrain such as urban environments, unfavorable weather patterns, and highly-trafficked sea lanes, when confronting U.S. forces. They will look to take advantage of strategic, operational, and tactical choke points for focusing their efforts. Complex terrain fosters highly decentralized and often unconnected events that detract from a conventional force's ability to concentrate its efforts.

Innovative Use of Emerging Technologies

Wide-ranging technological advances around the world will ultimately result in new capabilities and new weapons. In many cases – such as the Manhattan Project – the weapon itself is the driver behind the technology. Nano technology and bio-engineering are areas being pursued by many state and substate actors around the world -- most with the best of intentions. The future joint force commander can assume that as technology matures, elements of that technology will find a military application. While the continued development of existing WMD technologies and the emergence of new technologies are in themselves a future threat, the ability to combine several technologies can make this threat far more difficult to anticipate and control. Consider nano devices as a vector for the spread of a bio-engineered virus; a chemical agent that manipulates the effects of an electronic or radio frequency device; or a biological agent that remains dormant until activated by some form of electromagnetic energy.

Cognitive Campaigns

The cognitive domain has been defined as the area where "perceptions, awareness, beliefs, and values reside and where, as a result of sense-making, decisions are made."[62] Skillfully orchestrated actions in this domain may be central to future victory for both ourselves and for our adversaries. In the future, previously separate information streams will merge as increasingly dense information webs link adversaries, friends, citizens, and neutrals. Cognitive campaigns will likely be conduced with the goal of gaining the support of the nation, the world, and the local populace in the operational area while draining the will of the adversary. Goals of the cognitive campaign may be the driving factor for all other complementary operations—including political, military, economic or other activities that affect or influence the will of the adversary.

---

[62] David S. Alberts and Richard E. Hayes *Power to the Edge: Command and Control in the Information Age*. (CCRP: Washington, D.C.) p. 113.

Lawfare

While the United States is still obligated to respect and adhere to internationally accepted "laws of war" and legally binding treaties to which it is a signatory, there is increasing pressure to apply considerations appropriate for civil law to military operations and functions. The notion of "proportionality" dominates discussions regarding the appropriate level of military response. This idea addresses the American public's notion of fairness. Just as every crime does not deserve the death penalty, so every military threat does not require full-scale military retaliation. Although this notion has no basis in military theory, the U.S. must nevertheless be prepared to consider proportionality and the likely public perception of the appropriateness of a military response to a threat. In varying degrees all elements of national power (diplomatic, informational, military, and economic) must be considered and brought to bear in the correct proportions depending on the situation.

Other more specific legal issues are being raised regularly. In many cases individuals or non government organizations are challenging the military's relative freedom to wage war according to common military notions of necessity and of historical norms found in international law and existing treaties. For example, it has been suggested that enemy detainees and prisoners of war should be afforded the same legal protections enjoyed by U.S. citizens and residents. Regardless of the outcome of any single case, the U.S. must accept the fact that it will be continually challenged with regard to the legality of the war it is waging at the time and the ways and means by which it is conducted. Irrespective of the legality of policies, American political and military leaders must be sensitive to public perception.

4.5 Command

Command is the logic that translates the application of force into desired strategic outcomes. Command is the unifying vision and direction that directs a force to develop requisite knowledge of the adversary and to direct the application of force from one's base across terrain. It includes the authority and responsibility for effectively using available resources and for planning the employment of, organizing, directing, coordinating, and controlling forces for the accomplishment of assigned missions.[63]

- The Art and Science of War
- Pervasiveness and Influence of Networks
- Interaction of Military and Nonmilitary Domains

The Art and Science of War

Potential U.S. opponents understand a U.S way of war that is often focused on the "science" and "engineering" of military operations. Our cultural tendency towards a mechanistic view of war and its reduction to a relatively technical targeting exercise will encourage a competent adversary to rely on strategic and operational artistry to defeat the United States. They will understand the essential human nature of war, and will play to historical instincts of fear and honor that the West often believes it has left behind. Despite extensive experience in counterinsurgency, counterterrorist, and stability operations in Iraq and Afghanistan, the perception of potential opponents is likely to remain fixed on the idea that the U.S. is focused on technical notions of information dominance, speed, precision,

---

[63] Joint Publication 1-02. *Department of Defense Dictionary of Military and Related Terms.* 12 April 2001, as amended through 17 October 2007.

standoff technology, and dependence on air superiority to achieve overwhelming power against conventional force opponents. In their view, American confidence in the technical aspects of war has led to less emphasis on the political foundations of war, in planning for a viable political end state, and in matching national means to this end state. The implications of this foreign perception will be adversaries that are more willing and able to fight in the cultural and political domains. Adversary strategic and operational design will attempt to balance regional requirements to engage or even dominate neighbors, while simultaneously recognizing the need to shaping U.S. perception and engagement, while preparing for conflict with U.S. forces.

Despite popular notions that technology allows a world of bloodless and "humane" warfare, the future battlespace will not be a sterile, non-lethal world of robotic systems and point-and-click warriors. To the contrary, the future will require operational art that understands that integrated close combat will be much more episodic, dynamic, lethal, and unpredictable. For the whole of the joint force, it will be more intense, with increased tempo, and wider in scope. Future combat will also have greater psychological and emotional impact – and the increased power and influence placed in the hands of every member of the force will have dramatic and sometimes global effects. This future operational artistry will require greater teamwork at all levels across the entire joint force and will place significant demands on individual and unit discipline. Integrated, close combat will require mature leaders—mentally and physically tough—with superb cognitive and reasoning skills who are masters of tactical warfighting.

Pervasiveness and Influence of Networks

Cultural, organizational, and technological networks will be central to future methods of warfare. Networks are important because they increase the density of linkages and relationships among entities, actors, or systems. In the future, very few elements will be disconnected or unrelated to the wider world, and for this reason, activities will reverberate throughout the system, sometimes with unanticipated or detrimental effects. The network, as a conduit for information, influence, and collaboration allows adversaries new ways to affect the joint force.

The network offers new ways of maneuvering, in which commanders must think about how to move data, information, and knowledge. Military success in an environment of pervasive networking will go to commanders who encourages decentralized decision making at the lowest feasible levels and develop organizational constructs capable of self-synchronizing and self-adjusting swiftly as new conditions emerge. The more networked and less hierarchical environment will require personal and organizational command philosophy in which decisions are decentralized at the edge of the network.

Commanders must have the ability to assimilate a vast flood of intelligence data into a coherent, systemic understanding of adversary capabilities. This systemic understanding is about much more than simply placing weapons ever more effectively on target. It is about understanding exactly how an adversaries war-making potential is assembled. This change is characterized by the idea that any organization (state or its military forces, for example) operating at large scales is inherently "systemic" in nature, and relies on a dense web of linkages and relationships among a diverse array of constitutive elements to accomplish any purposeful action.

Interaction of Military and Nonmilitary Domains

Evolving U.S. joint operations doctrine posits a national-level campaign that focuses national capabilities—diplomatic, economic, information, and military—toward averting, deterring, and if necessary winning future conflicts. Once engaged, the United States must consider the political, economic, legal, military, and territorial aspects of the adversary's capability. In complex environments, multiple interactions constantly occur and effects the specific consequences of military activities will reverberate across each of these domains – and sometimes other unanticipated ones. The broader military implication is that the joint force commander must be prepared to engage with complexity, to understand that simple inputs to a system may have a number of effects, and to consider this complexity when planning and making decisions. Of particular importance will be the interaction of the military defense of the homeland and the multitude of civilian agencies, boards, authorities, corporations and governments responsible for running, maintaining, and protecting critical systems. Complex economies and societies like that of the United States' may be particularly susceptible to disruption due to the complex interactions and interdependencies of infrastructure and other elements of national power, complex distribution systems, the vagaries of globalization, and the ubiquitous nature of data, information, and knowledge.

Chapter 5: Conclusion

"War and warfare do not always change in an evolutionary or linear fashion. Surprise is not merely possible or even probable - it is certain."

~Colin Gray

With the above quote by Professor Gray in mind, we will, nonetheless attempt to narrow the nearly infinite range of future possibilities and humbly propose a small set of urgent joint force problem statements that attempt to capture, in a concise and understandable way, issues that the concept development and experimentation community throughout the Department of Defense and United States Government may wish to attempt to explore over the coming years in anticipation of the future operating environments that await. This list is neither static, nor self-contained and the JOE will continually solicit problem statements (as well as trends, variables, challenges, and military implications) from the futures community for inclusion in this document.

5.1 Notional Joint Force Problems

Problem 1

The first military problem defined by the trends and challenges in the JOE is the emergence of small, globally networked and distributed groups which conduct operations that:

- are supplied and based within the cultural and human terrain of societies from which they use as base;
- apply force in urban and complex terrain;
- attack U.S. forces and citizens with weapons with standoff from U.S. forces in both space (mortars, rocket-propelled grenades) and in time (improvised explosive devices), or by closing with them through suicide attacks and;
- Augment their tactical operations and protect their forces through the use of global media, transnational NGO's and the domestic legal systems of the United States and other free nations.

This combination of capabilities may deny the joint force the ability to identify a center of gravity, apply force to neutralize or destroy their forces locally and globally.

Problem 2

A second problem derived from the JOE is the emergence of several great powers that are armed with conventional military capabilities that may mirror the capabilities of U.S. forces – or even surpass them in some niche areas. These powers will have the capabilities to reach into space and cyberspace, and may be able to challenge the U.S. for dominance in these areas. They may develop information-enabled networked forces, and invest in naval forces with embarked stealthy air and undersea capabilities. Emerging great powers will seek to project power farther from their borders and may develop expeditionary capabilities to secure energy sources and supplies of natural resources. Emerging great powers will also rely on unique capabilities or local technologies to press geographic and societal advantages and to attack perceived U.S. vulnerabilities, especially in the ability to project the joint force across the globe into areas relatively near to the adversary.

Problem 3

A third problem is how to fight regional nuclear powers that use nuclear forces as an umbrella to secure dominance over local states. The joint force will have to leverage new national deterrent strategies against these states while conducting combat operations against them as necessary. The joint force must be able to conduct offensive operations against these states and counter powerful anti-access capabilities that limit our ability to bring mass to bear against them.

Problem 4

A fourth problem is how to conduct operations within failed mega-cities. The joint force must understand the physical and human terrain of three-dimensional cities hundreds of square miles in size where little legal order exists. It must understand how to separate hostile elements from the larger population and to bring fires against adversaries entrenched in the complex physical, human, and cultural environment of the mega-city.

5.2 The Way Forward

Our intention with this version of the Joint Operating Environment is to create a common baseline understanding of the trends, shocks, challenges, and joint force implications among the concept development, experimentation, and planning communities throughout the Department of Defense. The purpose of this baseline understanding is to support the exploration and development of joint forces capable of ensuring that the nation and its citizens are safe and prosperous and that the global international environment remains hospitable to U.S. values and traditions. The challenge for the near-term future will be to develop capabilities to address these problems in an anticipatory manner – before they become threats. In the longer-term future, the challenge should be to understand how the joint force (and the nation) can leverage its own dominant and asymmetric advantages in such as way as to create insoluble military and strategic dilemmas for adversaries so that they avoid challenging the U.S. altogether or are swiftly defeated should they attempt to engage.